Beginning Blockchain

A Beginner's Guide to Building Blockchain Solutions

Bikramaditya Singhal
Gautam Dhameja
Priyansu Sekhar Panda

ess®

Beginning Blockchain

Bikramaditya Singhal
Bangalore, Karnataka, India

Gautam Dhameja
Berlin, Berlin, Germany

Priyansu Sekhar Panda
Bangalore, Karnataka, India

ISBN-13 (pbk): 978-1-4842-3443-3
ISBN-13 (electronic): 978-1-4842-3444-0
https://doi.org/10.1007/978-1-4842-3444-0

Library of Congress Control Number: 2018945613

Managing Director, Apress Media LLC: Welmoed Spahr
Acquisitions Editor: Nikhil Karkal
Development Editor: Laura Berendson
Coordinating Editor: Divya Modi

Cover designed by eStudioCalamar

Cover image designed by Freepik (www.freepik.com)

Distributed to the book trade worldwide by Springer Science+Business Media New York, 233 Spring Street, 6th Floor, New York, NY 10013. Phone 1-800-SPRINGER, fax (201) 348-4505, e-mail orders-ny@springer-sbm.com, or visit www.springeronline.com. Apress Media, LLC is a California LLC and the sole member (owner) is Springer Science + Business Media Finance Inc (SSBM Finance Inc). SSBM Finance Inc is a **Delaware** corporation.

For information on translations, please e-mail rights@apress.com, or visit www.apress.com/rights-permissions.

Apress titles may be purchased in bulk for academic, corporate, or promotional use. eBook versions and licenses are also available for most titles. For more information, reference our Print and eBook Bulk Sales web page at www.apress.com/bulk-sales.

Any source code or other supplementary material referenced by the author in this book is available to readers on GitHub via the book's product page, located at www.apress.com/978-1-4842-3443-3. For more detailed information, please visit www.apress.com/source-code.

Printed on acid-free paper

Table of Contents

About the Authors ix

About the Technical Reviewer xi

Acknowledgments xiii

Introduction xv

Chapter 1: Introduction to Blockchain 1

- Backstory of Blockchain 2
- What is Blockchain? 4
- Centralized vs. Decentralized Systems 11
  - Centralized Systems 14
  - Decentralized Systems 15
- Layers of Blockchain 17
  - Application Layer 19
  - Execution Layer 20
  - Semantic Layer 20
  - Propagation Layer 21
  - Consensus Layer 22
- Why is Blockchain Important? 23
  - Limitations of Centralized Systems 23
  - Blockchain Adoption So Far 24
- Blockchain Uses and Use Cases 26
- Summary 28
- References 29

Chapter 2: How Blockchain Works 31
Laying the Blockchain Foundation 32
Cryptography..34
Symmetric Key Cryptography..37
Cryptographic Hash Functions 55
MAC and HMAC..76
Asymmetric Key Cryptography 78
Diffie-Hellman Key Exchange 98
Symmetric vs. Asymmetric Key Cryptography 102
Game Theory..104
Nash Equilibrium 107
Prisoner's Dilemma 108
Byzantine Generals' Problem..110
Zero-Sum Games..112
Why to Study Game Theory 113
Computer Science Engineering..114
The Blockchain 114
Merkle Trees 117
Putting It All Together..122
Properties of Blockchain Solutions..124
Blockchain Transactions..127
Distributed Consensus Mechanisms 130
Blockchain Applications..135
Scaling Blockchain..139
Off-Chain Computation..140
Sharding Blockchain State 143
Summary..145
References..146

Chapter 3: How Bitcoin Works.....149
The History of Money.....150
Dawn of Bitcoin.....153
What Is Bitcoin?.....154
Working with Bitcoins.....157
The Bitcoin Blockchain.....159
Block Structure.....161
The Genesis Block.....169
The Bitcoin Network.....172
Network Discovery for a New Node.....174
Bitcoin Transactions.....179
Consensus and Block Mining.....184
Block Propagation.....193
Putting It all Together.....195
Bitcoin Scripts.....195
Bitcoin Transactions Revisited.....196
Scripts.....204
Full Nodes vs. SPVs.....209
Full Nodes.....209
SPVs.....210
Bitcoin Wallets.....212
Summary.....216
References.....216

Chapter 4: How Ethereum Works.....219
From Bitcoin to Ethereum.....220
Ethereum as a Next-Gen Blockchain.....221
Design Philosophy of Ethereum.....223

Enter the Ethereum Blockchain....224
Ethereum Blockchain....225
Ethereum Accounts....228
Trie Usage....236
Merkle Patricia Tree....237
RLP Encoding....239
Ethereum Transaction and Message Structure....240
Ethereum State Transaction Function....245
Gas and Transaction Cost....248
Ethereum Smart Contracts....253
Contract Creation....256
Ethereum Virtual Machine and Code Execution....257
Ethereum Ecosystem....263
Swarm....264
Whisper....264
DApp....264
Development Components....265
Summary....265
References....266

Chapter 5: Blockchain Application Development 267

Decentralized Applications....267
Blockchain Application Development....269
Libraries and Tools....270
Interacting with the Bitcoin Blockchain....272
Setup and Initialize the bitcoinjs Library in a *node.js* Application....273
Create Keypairs for the Sender and Receiver....274
Get Test Bitcoins in the Sender's Wallet....275

Get the Sender's Unspent Outputs 276

Prepare Bitcoin Transaction 278

Sign Transaction Inputs 280

Create Transaction Hex 280

Broadcast Transaction to the Network 281

Interacting Programmatically with Ethereum—Sending Transactions 283

Set Up Library and Connection 284

Set Up Ethereum Accounts 285

Get Test Ether in Sender's Account 286

Prepare Ethereum Transaction 287

Sign Transaction 288

Send Transaction to the Ethereum Network 290

Interacting Programmatically with Ethereum—Creating a Smart Contract 292

Prerequisites 292

Program the Smart Contract 293

Compile Contract and Get Details 297

Deploy Contract to Ethereum Network 302

Interacting Programmatically with Ethereum—Executing Smart Contract Functions 307

Get Reference to the Smart Contract 308

Execute Smart Contract Function 309

Blockchain Concepts Revisited 312

Public vs. Private Blockchains 313

Decentralized Application Architecture 314

Public Nodes vs. Self-Hosted Nodes 315

Decentralized Applications and Servers 316

Summary 317

References 317

Chapter 6: Building an Ethereum DApp 319

The DApp 319

Setting Up a Private Ethereum Network 321

Install go-ethereum (*geth*) 321

Create *geth* Data Directory 322

Create a *geth* Account 323

Create *genesis.json* Configuration File 324

Run the First Node of the Private Network 325

Run the Second Node of the Network 329

Creating the Smart Contract 334

Deploying the Smart Contract 344

Setting up *web3* Library and Connection 345

Deploy the Contract to the Private Network 345

Client Application 359

Summary 375

References 375

Index 377

About the Authors

Bikramaditya Singhal is a Blockchain expert and AI practitioner with experience working in multiple industries. He is proficient in Blockchain, Bitcoin, Ethereum, Hyperledger, cryptography, cyber security, and data science. He has extensive experience in training and consulting on Blockchain and has designed many Blockchain solutions. He worked with companies such as WISeKey, Tech Mahindra, Microsoft India, Broadridge, and Chelsio Communications, and he also cofounded a company named Mund Consulting that focuses on big data analytics and artificial intelligence. He is an active speaker at various conferences, summits, and meetups. He has also authored a book entitled *Spark for Data Science.*

Gautam Dhameja is a Blockchain application consultant based out of Berlin, Germany. For most of this decade, he has been developing and delivering enterprise software including Web and Mobile apps, Cloud-based hyper-scale IoT solutions, and more recently, Blockchain-based decentralized applications (DApps). He possesses a deep understanding of the decentralized stack, cloud solutions architecture, and object-oriented design. His areas of expertise include Blockchain, Cloud-based enterprise solutions, Internet of Things, distributed systems, and native and hybrid mobile apps. He is also an active blogger and regularly speaks at tech conferences and events.

Priyansu Sekhar Panda is a research engineer at Underwriters Laboratories, Bangalore, India. He has worked with other IT companies such as Broadridge, Infosys Limited, and Tech Mahindra. His areas of expertise include Blockchain, Bitcoin, Ethereum, Hyperledger, game theory, IoT, and artificial intelligence. Priyansu's current research is on building next-gen applications leveraging Blockchain, IoT, and AI. His major research interests include building Decentralized Autonomous Organizations (DAO), and the security, scalability, and consensus of Blockchains.

About the Technical Reviewer

Navin K Manaswi has been developing AI solutions/products with the use of cutting-edge technologies and sciences related to artificial intelligence for many years. Having worked for consulting companies in Malaysia, Singapore, and the Dubai Smart City project, he has developed a rare skill of delivering end-to-end artificial intelligence solutions. He built solutions for video intelligence, document intelligence, and human-like chatbots in his own company. Currently, he solves B2B problems in verticals of healthcare, enterprise applications, industrial IoT, and retail in the Symphony AI incubator as Deep Learning-AI Architect. Through this book, he wants to democratize cognitive computing and services for everyone, especially developers, data scientists, software engineers, database engineers, data analysts, and CXOs.

Acknowledgments

We'd like to thank Nikhil and Divya for their cooperation and support all through and many thanks to Navin for his thorough technical review of this book. We also thank all who have directly or indirectly contributed to this book.

Introduction

Beginning Blockchain is a book for those willing to learn about the technical fundamentals of Blockchain, practical implications, and hands-on development aspects of Blockchain applications. Adequate history, background, and theoretical aspects are covered to help you build a solid foundation for your Blockchain journey, and appropriate development aspects are covered with coding examples to help you jumpstart on Blockchain assignments. The first chapter introduces you to the Blockchain world and sets the context. The second chapter dives deeper into the core components of Blockchain. The third chapter is focused on Bitcoin's technical concepts so what was discussed in the second chapter could be demonstrated with Bitcoin as a cryptocurrency use case of Blockchain. The fourth chapter is dedicated to the Ethereum Blockchain in an effort to demonstrate how Blockchain could be programmed for many more use cases and not limited to just cryptocurrencies. The fifth chapter is where you get the hang of Blockchain development with examples on basic transactions in both Bitcoin and Ethereum. The sixth chapter, as the final chapter, demonstrates the end-to-end development of a decentralized application (DApp). By the end of this chapter, you will have been equipped with enough tools and techniques to address many real-world business problems with applicable Blockchain solutions. Start your journey toward limitless possibilities.

CHAPTER 1

Introduction to Blockchain

Blockchain is the new wave of disruption that has already started to redesign business, social and political interactions, and any other way of value exchange. Again, it is not just a change, but a rapid phenomenon that is already in motion. As of this writing, more than 40 top financial institutions and many different firms across industries have started to explore blockchain to lower transaction cost, speed up transaction time, reduce the risk of fraud, and eliminate the middleman or intermediary services. Some are trying to reimagine legacy systems and services to take them to a new level and also come up with new kinds of service offerings.

We will cover blockchain in greater detail throughout the book. You can follow through the chapters in the order presented if you are new to blockchain or pick only the ones relevant to you. This chapter explains what blockchain is all about, how it has evolved, and its importance in today's world with some uses and use cases. It gives an outside-in perspective to you to be able to delve deeper into blockchain.

B. Singhal et al., *Beginning Blockchain*, https://doi.org/10.1007/978-1-4842-3444-0_1

Backstory of Blockchain

One of the first known digital disruptions that laid the foundation of the Internet was TCP/IP (Transmission Control Protocol/Internet Protocol) back in the 1970s. Prior to TCP/IP, it was the era of circuit switching, which required dedicated connection between two parties for communication to happen. TCP/IP came up with its packet switching design, which was more open and peer-to-peer with no need to preestablish a dedicated line between parties.

When the Internet was made accessible to the public through the World Wide Web (WWW) in the early 1990s, it was supposed to be more open and peer-to- peer. This is because it was built atop the open and decentralized TCP/IP. When any new technology, especially the revolutionary ones, hits the market, either they fade away on their own, or they create such an impact that they become the accepted norm. People adapted to the WWW revolution and leveraged the benefits it had to offer in every possible way. As a result, the World Wide Web started shaping up in a way that might not have been the exact way it was imagined. It could have been more open, more accessible, and more peer-to-peer. Many new technologies and businesses started to build on top of it and it became what it is today–more centralized. Slowly and gradually, people get used to what technology offers. People are just fine if an international transaction takes days to settle, or it is too expensive, or it is less reliable.

Let us take a closer look at the banking system and its evolution. Starting from the olden days of barter system till fiat currencies, there was no real difference between a transaction and its settlement because they were not two separate entities. As an example, if Alice had to pay $10 to Bob, she would just hand over a $10 note to Bob and the transaction would just get settled there. No bank was needed to debit $10 from Alice's account and credit the same to Bob's account or to serve as a system of trust to ensure Alice does not cheat Bob. However, transacting directly with someone who is physically not present close by was difficult. So, banking systems evolved with many more service offerings and enabled transactions from every corner of the world. With the help of the Internet, geography was no more a limitation and banking became easier than ever. Not just banking for that matter: the Internet facilitated many different kinds of value exchange over the web.

Technology enabled someone from India to make a monetary transaction with someone in the United Kingdom, but with some cost. It takes days to settle such transactions and is expensive as well. A bank was always needed to impose trust and ensure security for such transactions between two or more parties. What if technology could enable trust and security without these intermediary and centralized systems? Somehow, this part (of technology imposing trust) was missing all through, which resulted in development of centralized systems such as banks, escrow services, clearing houses, registrars and many other such institutions. Blockchain proves to be that missing piece of the Internet revolution puzzle that facilitates a trustless system in a cryptographically secured way.

Satoshi Nakamoto, the pseudonymous name by which the world knows him, must have felt that the monetary systems were not touched by the technological revolution since the 1980s. Banks formed the centralized institutions that maintained the transaction records, governed interactions, enforced trust and security, and regulated the whole system. The whole of commerce relies on these financial institutions, which serve as the trusted third parties to process payments. The mediation of financial institutions increases cost and time to settle a transaction, and also limits the transaction sizes. The mediation was needed to settle disputes, but that meant that completely nonreversible transaction was never possible. This resulted in a situation where trust was needed for someone to transact with another. Certainly, this bureaucratic system had to change to keep up with the economy's expected digital transformation. So, Satoshi invented a cryptocurrency called Bitcoin that was enabled by the underlying technology– blockchain. Bitcoin is just one monetary use case of blockchain that addresses the inherent weakness of trust-based models. We will delve deeper into the fundamentals of both Bitcoins and blockchain in this book.

What is Blockchain?

The Internet has revolutionized many aspects of life, society, and business. However, we learned in the previous section that how people and organizations execute transactions with one another has not changed much in the past couple of decades. Blockchain is believed to be the component that completes the Internet puzzle and makes it more open, more accessible, and more reliable.

To understand blockchain, you have to understand it from both a business perspective and technical perspective. Let us first understand it in a business transaction context to get the "what" of it, and then look into the technicality to understand the "how" of it in the following chapters.

Blockchain is a system of records to transact value (not just money!) in a peer-to-peer fashion. What it means is that there is no need for a trusted intermediary such as banks, brokers, or other escrow services to serve as a trusted third party. For example, if Alice pays to Bob $10, why would it go through a bank? Take a look at Figure 1-1.

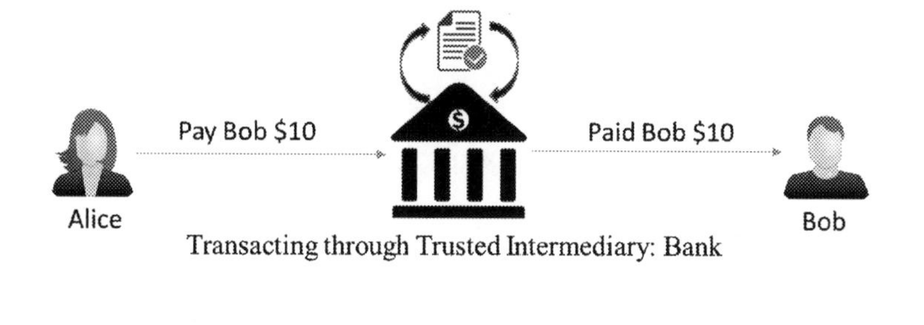

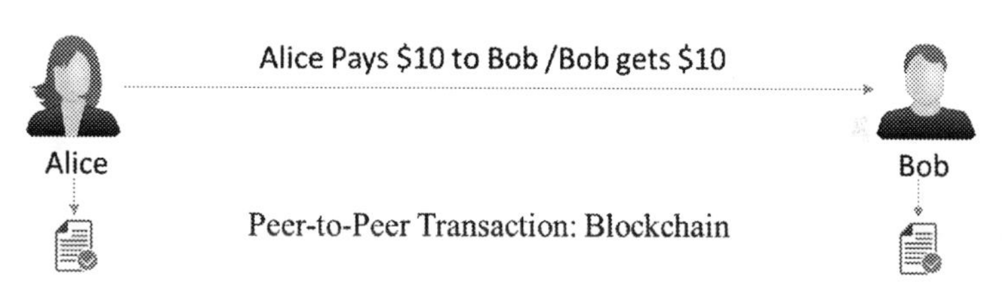

Figure 1-1. *Transaction through an intermediary vs. peer-to-peer transaction*

Let us look at a different example now. A typical stock transaction happens in seconds, but its settlement takes weeks. Is it desirable in this digital age? Certainly not! Figure 1-2 demonstrates the current situation.

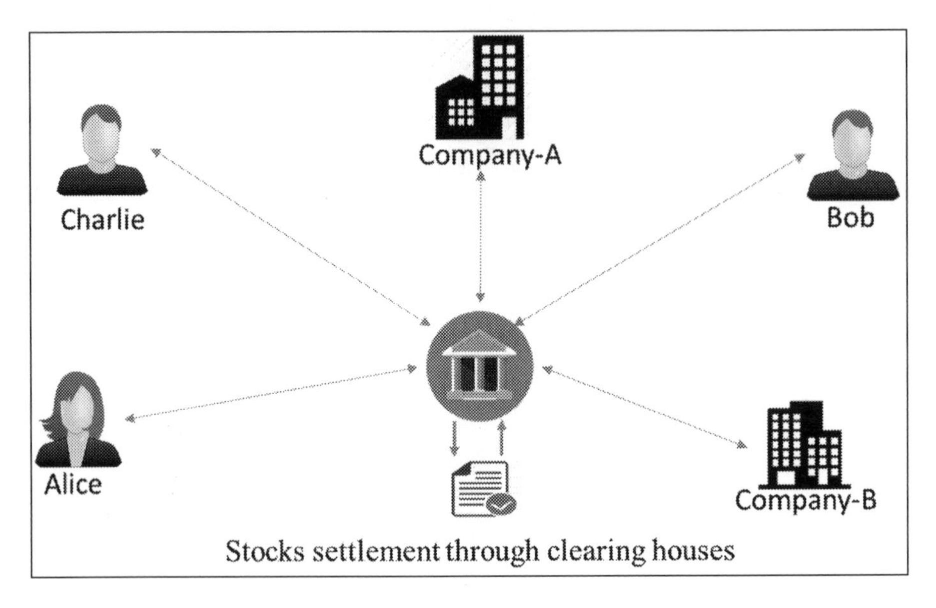

Figure 1-2. *Stocks trading through an intermediary clearing house*

If someone wants to buy some stocks from a company or a person, they can just directly buy it from them with instant settlement, with no need for brokers, clearing houses, or other financial institutions in between. A decentralized and peer-to-peer solution to such a situation can be represented as in Figure 1-3.

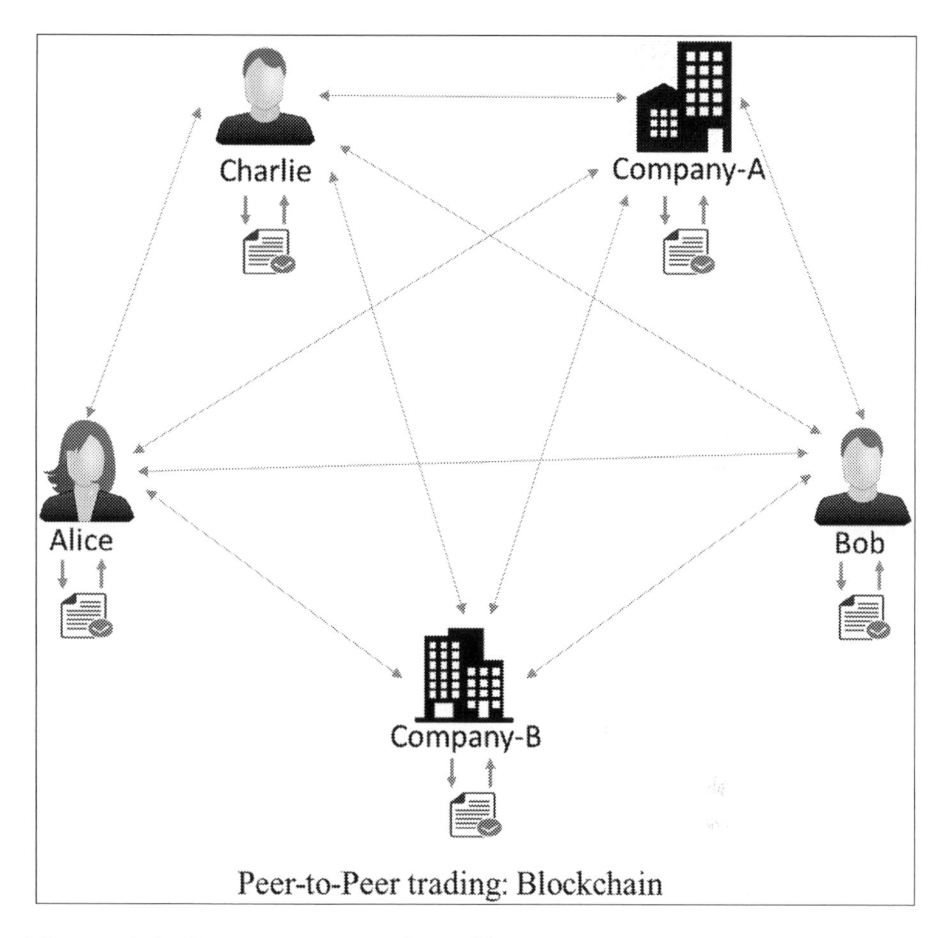

Figure 1-3. *Peer-to-peer stock trading*

Please note that transaction and settlement are not two different entities in a blockchain setting! The transactions are analogous to, say, fiat currency transactions where if someone pays another a $10 note, they do not own it anymore and that $10 note is physically transferred to the new owner.

Now that you understand blockchain from a functional perspective, at a high level, let us look into it a little technically, and the reason for naming it “blockchain” becomes clearer. We will see “What” it is technically and leave the “How” it works to Chapter 2.

Read the following statements and do not worry if the concepts do not fit together well for your complete understanding. You may want to revisit it, but be patient till you finish reading this book.

- Blockchain is a peer-to-peer system of transacting values with no trusted third parties in between.
- It is a shared, decentralized, and open ledger of transactions. This ledger database is replicated across a large number of nodes.
- This ledger database is an append-only database and cannot be changed or altered. It means that every entry is a permanent entry. Any new entry on it gets reflected on all copies of the databases hosted on different nodes.
- There is no need for trusted third parties to serve as intermediaries to verify, secure, and settle the transactions.
- It is another layer on top of the Internet and can coexist with other Internet technologies.
- Just the way TCP/IP was designed to achieve an open system, blockchain technology was designed to enable true decentralization. In an effort to do so, the creators of Bitcoin open-sourced it so it could inspire many decentralized applications.

A typical blockchain may look as shown in Figure 1-4.

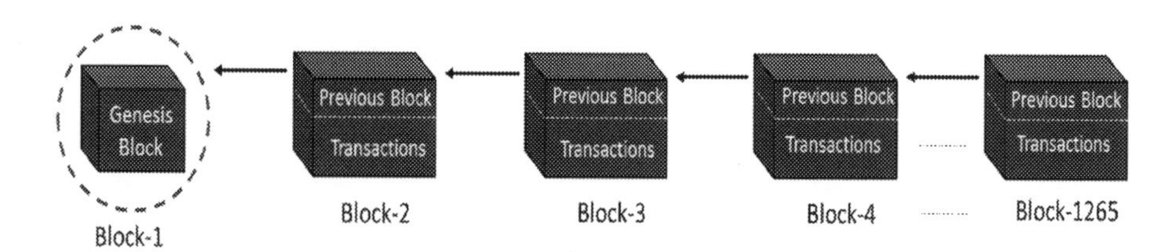

Figure 1-4. *The blockchain data structure*

Every node on the blockchain network has an identical copy of the blockchain shown in Figure 1-4, where every block is a collection of transactions, hence the name. As you can see, there are two major parts in every block. The "header" part links back to the previous block in the chain. What it means is that every block header contains the hash of the previous block so that no one can alter any transaction in the previous block. We will look into further details of this concept in the following chapters. The other part of a block is the "body content" that has a validated list of transactions, their amounts, the addresses of the parties involved, and some more details. So, given the latest block, it is feasible to access all the previous blocks in a blockchain.

Let us consider a practical example and see how the transactions take place and the ledger gets updated across the network, to see how this system works:

Assume that there are three candidates–Alice, Bob, and Charlie–who are doing some monetary transactions among each other on a blockchain network. Let us go through the transactions step by step to understand blockchain's open and decentralized features.

Step-1:

Let us assume that Alice had $50 with her, which is the genesis of all transactions and every node is aware of it, as shown in Figure 1-5.

Figure 1-5. *The genesis block*

Step-2:

Alice makes a transaction by paying $20 to Bob. Observe how the blockchain gets updated at each node, as shown in Figure 1-6.

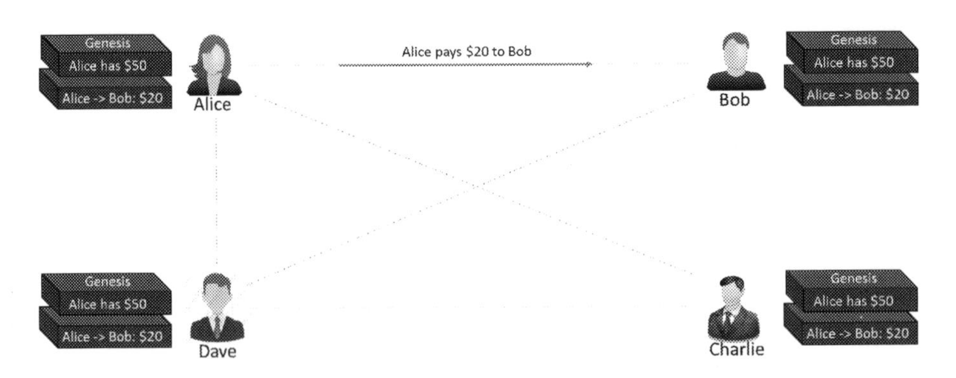

Figure 1-6. *The first transaction*

Step-3:

Bob makes another transaction by paying $10 to Charlie and the blockchain gets updated as shown in Figure 1-7.

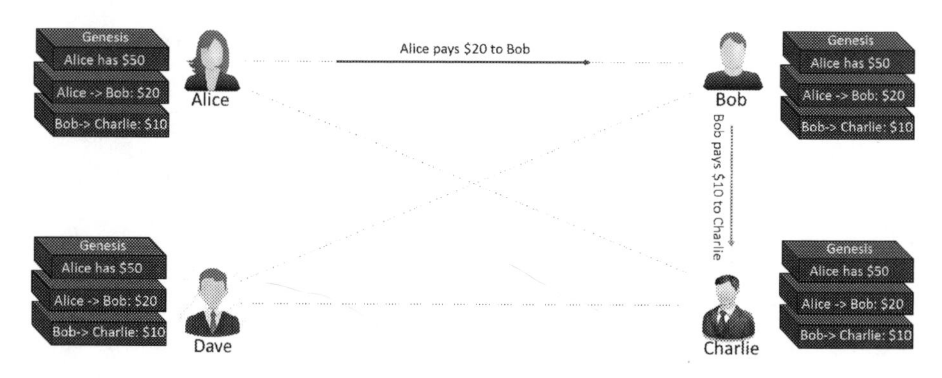

Figure 1-7. *The second transaction*

Please note that the transaction data in the blocks is immutable. All transactions are fully irreversible. Any change would result in a new transaction, which would get validated by all contributing nodes. Every node has its own copy of blockchain.

If there are many questions popping up in your mind, such as "What if Alice pays the same amount to Dave to double-spend the same amount, or what if she is making a payment without having enough funds in her account?," "How is the security ensured?," and so on, that is wonderful! We are going to cover those details in the following chapters.

Centralized vs. Decentralized Systems

The very reason we are looking into the debate on centralization vs. decentralization is because blockchain is designed to be decentralized, defying the centralized design. However, the terns *decentralized* and *centralized* are not always clear. They are very poorly defined and misleading in many places. The reason is that there is almost no system that is purely centralized or decentralized. Most of the concepts and examples in this section are inspired from the notes of Mr. Vitalik Buterin, the founder of Ethereum blockchain.

What is a distributed system then? Just so it does not mess with the current discussion, let us understand it first and take it off the list. Please note that whether a system is centralized or decentralized, it can still be distributed. A centralized distributed system is one in which there is, say, a master node responsible for breaking down the tasks or data and distribute the load across nodes. On the other hand, a decentralized distributed system is one where there is no "master" node as such and yet the computation may be distributed. Blockchain is one such example, and we will look at many diagrammatic representations of it later in this book. Figure 1-8 is a pictorial representation of how a centralized distributed system may look.

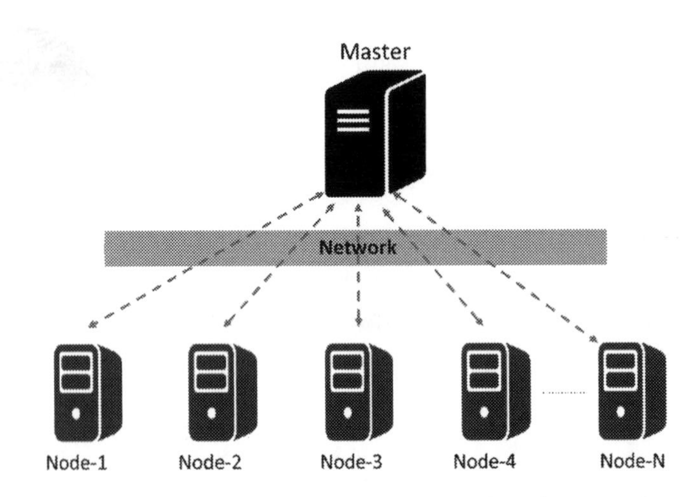

***Figure 1-8.** A distributed system with centralized control*

This representation is similar to Hadoop implementation, as an example. Though the computation is faster in such designs because of distributed computing, it also suffers from limitations due to centralization.

Let us resume our discussion on centralization vs. decentralization. It is extremely important to note that for a system to be centralized/decentralized is not just limited to the technical architecture. What we intend to say is that a system can be centralized or decentralized technically, but may not be so logically or politically. Let us take a look at these different perspectives to be able to design a system correctly based on the requirement:

Technical Architecture: A system can be centralized or decentralized from a technical architecture point of view. What we consider is how many physical computers (or nodes) are used to design a system, how many node failures it can sustain before the whole system goes down, etc.

Political perspective: This perspective indicates the control that an individual, or a group of people, or an organization as a whole has on a system. If the computers of the system are controlled by them, then the system is naturally centralized. However, if no specific individual or groups control the system and everyone has equal rights on the system, then it is a decentralized system in a political sense!

Logical perspective: A system can be logically centralized or decentralized based on how it appears to be, irrespective of whether it is centralized or decentralized technically or politically. An alternative analogy could be that if you vertically cut a system (say of computing devices) in half with every half having service providers and consumers, if they can operate as independent units they are decentralized and centralized otherwise.

All the aforementioned perspectives are crucial in designing a real-life system and calling it centralized or decentralized. Let us discuss some of the examples mixing these perspectives to clear up any confusion you might have:

- If you look at the corporates, they are architecturally centralized (one head office), they are politically centralized (governed by a CEO or the board), and they are logically centralized as well. (You can't really split them in halves.)

- Our language of communication is decentralized from every perspective—architecturally, politically, as well as logically. For two people to communicate with each other, in general, their language is neither politically influenced nor logically dependent on the language of communication of other people.
- The torrent systems such as BitTorrent are also decentralized from every perspective. Any node can be a provider or a consumer, so even if you cut the system into halves, it still sustains.
- The Content Delivery Network on the other hand is architecturally decentralized, logically also decentralized, but is politically centralized because it is owned by corporates. An example is Amazon CloudFront.
- Let us consider blockchain now. The objective of blockchain was to enable decentralization. So, it is architecturally decentralized by design. Also, it is decentralized from a political viewpoint, as nobody controls it. However, it is logically centralized, as there is one common agreed state and the whole system behaves like a single global computer.

Let us explore these terms separately and have a comparative view to be able to appreciate why blockchain is decentralized by design.

Centralized Systems

As the name suggests, a centralized system has a centralized control with all administrative authority. Such systems are easy to design, maintain, impose trust, and govern, but suffer from many inherent limitations, as follows:

- They have a central point of failure, so are less stable.
- They are more vulnerable to attack and hence less secured.
- Centralization of power can lead to unethical operations.
- Scalability is difficult most of the time.

A typical centralized system may appear as shown in Figure 1-9.

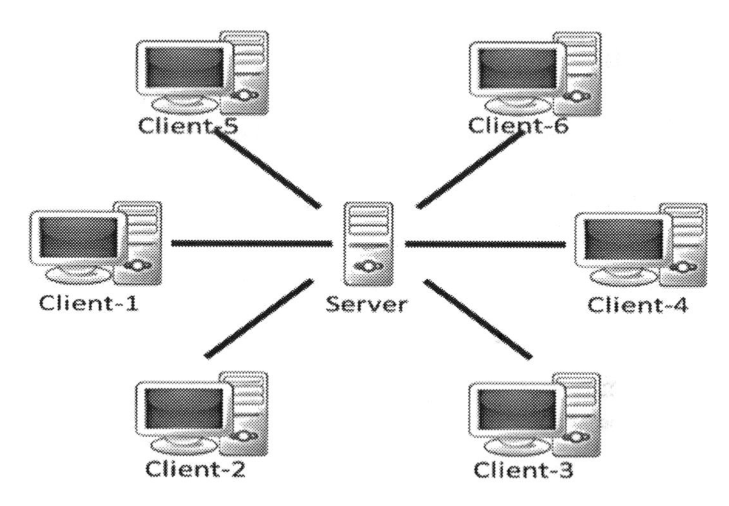

Figure 1-9. *A centralized system*

Decentralized Systems

As the name suggests, a decentralized system does not have a centralized control and every node has equal authority. Such systems are difficult to design, maintain, govern, or impose trust. However, they do not suffer from the limitations of conventional centralized systems. Decentralized systems offer the following advantages:

- They do not have a central point of failure, so more stable and fault tolerant

- Attack resistant, as no central point to easily attack and hence more secured
- Symmetric system with equal authority to all, so less scope of unethical operations and usually democratic in nature

A typical decentralized system may appear as shown in Figure 1-10.

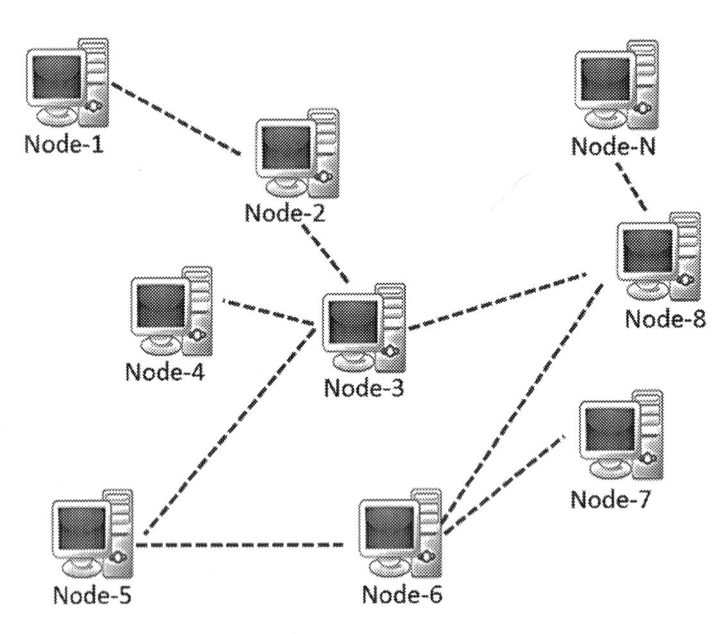

Figure 1-10. *A decentralized system*

Please note that a distributed system can also be decentralized. An example would be blockchain! However, unlike common distributed systems, the task is not subdivided and delegated to nodes, as there is no master who would do that in blockchain. The contributing nodes do not work on a portion of the work, rather, the interested nodes (or the ones chosen at random) perform the entire work. A typical decentralized and distributed system, which is effectively a peer-to-peer system, may appear as shown in Figure 1-11.

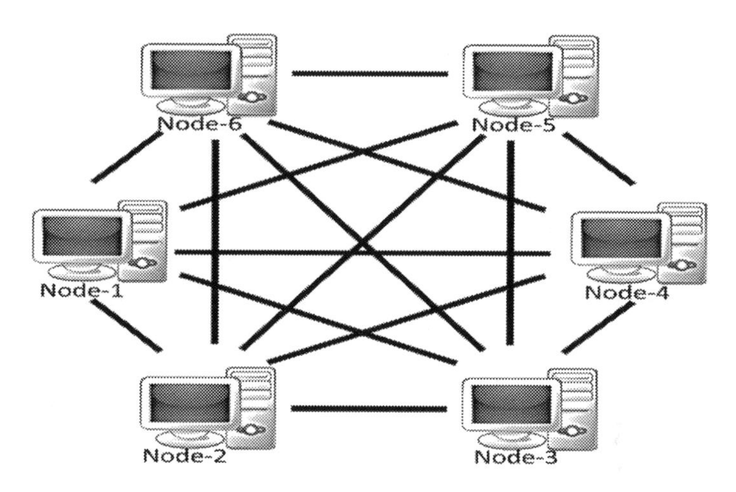

Figure 1-11. *A decentralized and peer-to-peer system*

Layers of Blockchain

As of this writing, the public blockchain variants such as Ethereum are in the process of maturing, and building complex applications on top of these blockchains may not be a good idea. Keep in mind that blockchain is never just a piece of technology, but a combination of business principles, economics, game theory, cryptography, and computer science engineering. Most of the real-world applications are quite complex in nature, and it is advisable to build blockchain solutions from the ground up.

The purpose of this section is only to provide you with a bird's eye view of various blockchain layers, and delve deeper into the core fundamentals in the following chapters. To start with, let us just recollect our basic understanding of the TCP/IP protocol stack. The layered approach in the TCP/IP stack is actually a standard to achieve an open system. Having abstraction layers not only helps in understanding the stack better, but also helps in building products that are compliant to the stack to achieve an open system. Also, having the layers abstract from each other makes the system more robust and easy to maintain. Any change to any of the layers doesn't impact the other layers. Again, the TCP/IP analogy is not to be confused with the blockchain layers. TCP/IP is a communication protocol that every Internet application uses, and so is blockchain.

Enter the blockchain. There are no agreed global standards yet that would clearly segregate the blockchain components into distinct layers. A layered heterogeneous architecture is needed, but for now that is still in the future. So, we will try to formulate blockchain layers to be able to understand the technology better and build a comparative analogy between hundreds of blockchain/Cryptocurrency variants out there in the market. Take a look at the high-level, layered representation of blockchain in Figure 1-12.

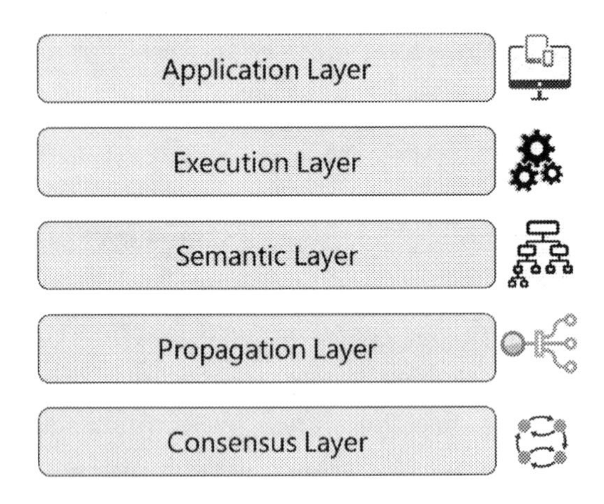

***Figure 1-12.** Various layers of blockchain*

You may be wondering why five layers and why not more granular layers, or fewer layers. Obviously, there cannot be too many or too few layers; it is going to be a trade-off driven among complexity, robustness, adaptability, etc., to name a few. The purpose again is not really to standardize blockchain technology, but to build a better understanding. Please keep in mind that all these layers are present on all the nodes.

In Chapter 6 of this book, we will be building a decentralized application from scratch and learning how blockchain functions on all these layers with a practical use case.

Application Layer

Because of the characteristics of blockchain, such as immutability of data, transparency among participants, resilience against adversarial attacks etc., there are multiple applications being built. Certain applications are just built in the application layer, taking for granted any available "flavor" of blockchain, and some applications are built in the application layer and are interwoven with other layers in blockchain. This is the reason the application layer should be considered a part of blockchain.

This is the layer where you code up the desired functionalities and make an application out of it for the end users. It usually involves a traditional tech stack for software development such as client-side programming constructs, scripting, APIs, development frameworks, etc. For the applications that treat blockchain as a backend, those applications might need to be hosted on some web servers and that might require web application development, server-side programming, and APIs, etc. Ideally, good blockchain applications do not have a client-server model, and there are no centralized servers that the clients access, which is just the way Bitcoin works.

You probably have heard or already learned about the off-chain networks. The idea is to build applications that wouldn't use blockchain for anything and everything, but use it wisely. In other words, this concept is to ensure that the heavy lifting is done at the application layer, or bulky storage requirements are taken care of off the chain so that the core blockchain is light and effective and the network traffic is not too much.

Execution Layer

The Execution Layer is where the executions of instructions ordered by the Application Layer take place on all the nodes in a blockchain network. The instructions could be simple instructions or a set of multiple instructions in the form of a *smart contract*. In either case, a program or a script needs to be executed to ensure the correct execution of the transaction. All the nodes in a blockchain network have to execute the programs/scripts independently. Deterministic execution of programs/scripts on the same set of inputs and conditions always produces the same output on all the nodes, which helps avoid inconsistencies.

In the case of Bitcoins, these are simple scripts that are not Turing-complete and allow only a few set of instructions. Ethereum and Hyperledger, on the other hand, allow complex executions. Ethereum's code or its smart contracts written in solidity gets compiled to Bytecode or Machine Code that gets executed on its own Ethereum Virtual Machine. Hyperledger has a much simpler approach for its chaincode smart contracts. It supports running of compiled machine codes inside docker images, and supports multiple high-level lanuages such as Java and Go.

Semantic Layer

The Semantic Layer is a logical layer because there is an orderliness in the transactions and blocks. A transaction, whether valid or invalid, has a set of instructions that gets through the Execution Layer but gets validated in the Semantic Layer. If it is Bitcoin, then whether one is spending a legitimate transaction, whether it is a double-spend attack, whether one is authorized to make this transaction, etc., are validated in this layer. You will learn in the following chapters that Bitcoins are actually present as transactions that represent the system state. To be able to spend a Bitcoin, you have to consume one or more previous transactions and there is no notion of *Accounts*. This means that when someone makes a transaction, they use one of the previous transactions where they had received at least the amount they are spending now. This transaction must be validated by all the nodes by traversing previous transactions to see if it is a legitimate transaction. Ethereum, on the other hand, has the system of *Accounts*. This means that the account of the one making the transaction and that of the one receiving it both get updated.

In this layer, the rules of the system can be defined, such as data models and structures. There could be situations that are a little more complex compared with simple transactions. Complex instruction sets are often coded into smart contracts. The system's state gets updated when a smart contract is invoked upon receiving a transaction. A smart contract is a special type of account that has executable code and private states. A block usually contains a bunch of transactions and some smart contracts. The data structures such as the Merkle tree are defined in this layer with the Merkle root in the block header to maintain a relation between the block headers and the set of transactions in a block (usually Key-Value storage on disk). Also, the data models, storage modes, in-memory/disk based processing, etc. can be defined in this logical layer.

Apart from the aforementioned, it is the semantic layer that defines how the blocks are linked with each other. Every block in a blockchain contains the hash of the previous block, all the way to the genesis block. Though the final state of the blockchain is achieved by the contributions from all the layers, the linking of blocks with each other needs to be defined in this layer. Depending on the use case, you might want to code up an additional functionality in this layer.

Propagation Layer

The previous layers were more of an individual phenomenon: not much coordination with other nodes in the system. The Propagation Layer is the peer-to-peer communication layer that allows the nodes to discover each other, and talk and sync with each other with respect to the current state of the network. When a transaction is made, we know that it gets broadcast to the entire network. Similarly, when a node wants to propose a valid block, it gets immediately propagated to the entire network so that other nodes could build on it, considering it as the latest block. So, transaction/block propagation in the network is defined in this layer, which ensures stability of the whole network. By design, most of the blockchains are designed such that they forward a transaction/block immediately to all the nodes they are directly connected to, when they get to know of a new transaction/block.

In the asynchronous Internet network, there are often latency issues for transaction or block propagation. Some propagations occur within seconds and some take more time, depending on the capacity of the nodes, network bandwidth, and a few more factors.

Consensus Layer

The Consensus Layer is usually the base layer for most of the blockchain systems. The primary purpose of this layer is to get all the nodes to agree on one consistent state of the ledger. There could be different ways of achieving consensus among the nodes, depending on the use case. Safety and security of the blockchain is accertained in this layer. In Bitcoin or Ethereum, the consensus is achieved through proper incentive techniques called "mining." For a public blockchain to be self-sustainable, there has to be some sort of incentivization mechanisms that not only helps in keeping the network alive, but also enforces consensus. Bitcoin and Ethereum use a Proof of Work (PoW) consensus mechanism to randomly select a node that can propose a block. Once that block is proposed and propagated to all the nodes, they check to see if it a valid block with all legitimate transactions and that the PoW puzzle was solved properly; they add this block to their own copy of blockchain and build further on it. There are many different variants of consensus protocols such as Proof of Stake (PoS), deligated PoS (dPoS), Practical Byzantine Fault Tolerance (PBFT), etc., which we will cover in great detail in the following chapters.

Why is Blockchain Important?

We looked at the design aspects of centralized and decentralized systems and got some idea of the technical benefits of decentralized systems over centralized ones. We also learned about different layers of blockchain. Blockchain, being a decentralized peer-to-peer system, has some inherent benefits and complexities. Keep in mind that it is not a silver bullet that can address all the problem areas in the world, but there are specific cases where it is the need of the hour. There are also scenarios where blockchainizing the existing solution makes it more robust, transparent, and secured. However, it can as well lead to disaster if not done the right way! Let us now keep a business and functional perspective in mind and analyze blockchain.

Limitations of Centralized Systems

If you take a quick glance at the software evolution landscape, you will see that many software solutions have a centralized design. The reason is not just because they are easy to develop and maintain, but because we are used to such a design to be able to trust the system. We always need a trusted third party who can assure we are not being cheated or becoming victims of a scam. Without a prior business relationship, it is difficult to trade with someone or even scale up. One would probably not do business with someone they have never known.

Let us take an example to understand it better. Today when we order something from Amazon, we feel safe and assured of the item's delivery. The producer of the item is someone and the buyer is someone else. Then what role is being played by Amazon here? It is there as an enabler functioning as a trusted intermediary, and also to take some cut of the transaction. The buyer trusts the seller where the trust relation is actually imposed by such trusted third parties. What blockchain proposes is that, in the modern digital era, we do not really need a third party in between to impose trust, and the technology has matured enough to handle it. In blockchain, trust is an inherent part of the network by default, which we will explore more in upcoming chapters.

Let us quickly learn a few downsides of a conventional centralized system:

- Trust issues
- Security issue
- Privacy issue—data sale privacy is being undermined
- Cost and time factor for transactions

Some of the advantages of decentralized systems over centralized systems could be:

- Elimination of intermediaries
- Easier and genuine verification of transactions
- Increased security with lower cost
- Greater transparency
- Decentralized and immutable

Blockchain Adoption So Far

Blockchain came along with Bitcoin, a digital cryptocurrency, in 2009 via a simple mailing list. Soon after it was launched, people could realize its true potential beyond just cryptocurrency. Some companies came up with different flavors of blockchain offerings such as Ethereum, Hyperledger, etc. Microsoft and IBM came up with SaaS (Software as a Service) offerings on their Azure and Bluemix cloud platforms, respectively. Different start-ups were formed, and many established companies took blockchain initiatives that focused on solving some business problems that were not easily solved before.

It is too late now to just say that blockchain has tremendous potential to disrupt almost every industry in some way or the other; the revolution has already started. It has hugely impacted the financial services market. It is difficult to name a global bank or finance entity not exploring blockchain. Apart from the financial market, initiatives have already been/are already being taken in areas such as media and entertainment, energy trading, prediction markets, retail chains, loyalty rewards systems, insurance, logistics and supply chains, medical records, and also government and military applications.

As of this writing, the current situation is such that many start-ups and companies are able to see how a blockchain-based system can really address some pain areas and become beneficial in many ways. However, designing the right kind of blockchain solution is quite challenging. There are some really great ideas for a blockchain-based product or solution, but it is equally difficult to either build them or implement them. There are some use cases that can only be built on a public blockchain. Designing a self-sustainable blockchain with a proper mining ecosystem is difficult, and when it comes to the existing public blockchains to build non-cryptocurrency applications there is none other than Ethereum. Whether a blockchain application is to be built in the Application Layer only and use the underlying layers as they are, or the application needs to be built from the ground up, is something difficult to decide. There are some technical challenges, too. Blockchain is still maturing, and it may take few more years for mainstream adoption. As of today, there are multiple propositions to address the scalability issues of blockchain. We will try to build a solid understanding on all these perspectives in this entire book. For now, let us see some of the specific uses and use cases in the following section.

Blockchain Uses and Use Cases

In this section, we will look at some of the initiatives that are already being taken across industries such as finance, insurance, banking, healthcare, government, supply chains, IoT (Internet of Things), and media and entertainment to name a few. The possibilities are limitless, however! A true sharing economy, which was difficult to achieve in centralized systems, is possible using blockchain technology (e.g., peer-to-peer versions of Uber, AirBNB). It is also possible to enable citizens to own their identity (Self-Sovereign Digital Identity) and monetize their own data using this technology. For now, let us take a look at some of the existing use cases.

- Any type of property or asset, whether physical or digital, such as laptops, mobile phones, diamonds, automobiles, real estate, e-registrations, digital files, etc. can be registered on blockchain. This can enable these asset transactions from one person to another, maintain the transaction log, and check validity or ownerships. Also, notary services, proof of existence, tailored insurance schemes, and many more such use cases can be developed.
- There are many financial use cases being developed on blockchain such as cross-border payments, share trading, loyalty and rewards system, Know Your Customer (KYC) among banks, etc. Initial Coin Offering (ICO) is one of the most trending use cases as of this writing. ICO is the best way of crowdsourcing today by using cryptocurrency as digital assets. A coin in an ICO can be thought of as a digital stock in an enterprise, which is very easy to buy and trade.

- Blockchain can be used to enable “The Wisdom of Crowds” to take the lead and shape businesses, economies, and various other national phenomena by using collective wisdom! Financial and economic forecasts based on the wisdom of crowds, decentralized prediction markets, decentralized voting, as well as stocks trading can be possible on blockchain.
- The process of determining music royalties has always been convoluted. The Internet-enabled music streaming services facilitated higher market penetration, but made the royalty determination more complex. This concern can pretty much be addressed by blockchain by maintaining a public ledger of music rights ownership information as well as authorised distribution of media content.
- This is the IoT era, with billions of IoT devices everywhere and many more to join the pool. A whole bunch of different makes, models, and communication protocols makes it difficult to have a centralized system to control the devices and provide a common data exchange platform. This is also an area where blockchain can be used to build a decentralized peer-to-peer system for the IoT devices to communicate with each other. ADEPT (Autonomous Decentralized Peer-To-Peer Telemetry) is a joint initiative from IBM and Samsung that has developed a platform that uses elements of the Bitcoin’s underlying design to build a distributed network of devices—a decentralized IOT. ADEPT uses three protocols: BitTorrent for file sharing, Ethereum for smart contracts, and TeleHash for peer-to-peer messaging in the platform. The IOTA foundation is another such initiative.

- In the government sectors as well, blockchain has gained momentum. There are use cases where technical decentralization is necessary, but politically should be governed by governments: land registration, vehicle registration and management, e-Voting, etc. are some of the active use cases. Supply chains are another area where there are some great use cases of blockchain. Supply chains have always been prone to disputes across the globe, as it was always difficult to maintain transparency in these systems.

Summary

In this chapter, we covered the evolution of blockchain, the history of it, what it is, the design benefits, and why it is so important with some relevant use cases. In this section, we will conclude with its game-changing offerings, in line with the technology revolution.

In the 1990s, mass adoption of the Internet changed the way people did business. It removed friction from creation and distribution of information. This paved the way for new markets, more opportunities, and possibilities. Similarly, blockchain is here today to take the Internet to a whole new level by removing friction along three key areas: Control, Trust, and Value.

Control: Blockchain enabled distribution of the control by making the system decentralized.

Trust: Blockchain is an immutable, tamper-resistant ledger. It gives a single, shared source of truth to all nodes, making the system trustless. What it means is that trust is no longer needed to transact with any unknown person or entity and is inherent by design.

Value: Blockchain enables exchange of value in any form. One can issue and transfer assets without central entities or intermediaries.

In Chapter 2, we will take a deep dive into the blockchain fundamentals.

References

The Meaning of Decentralization

Buterin, Vitalik, "The Meaning of Decentralization," *Medium*, https://medium.com/@VitalikButerin/the-meaning-of-decentralization-a0c92b76a274, February 6, 2017.

BlockChain Technology

Crosby, Michael; Nachiappan; Pattanayak, Pradhan; Verma, Sanjeev; Kalyanaraman, Vignesh, "BlockChain Technology: Beyond Bitcoin," Berkeley, CA: Sutardja Center for Entrepreneurship & Technology, University of California, http://scet.berkeley.edu/wp-content/uploads/BlockchainPaper.pdf, October 16, 2015.

Torpey, Kyle, "Eric Lombrozo: Bitcoin Needs Protocol Layers Similar to the Internet," *CoinJournal*, https://coinjournal.net/eric-lombrozo-bitcoin-needs-protocol-layers-similar-to-the-internet/, January 28, 2016.

Blockbench: A Framework for Analyzing Private blockchains

Dinh, Tien Tuan Anh; Wang, Ji; Chen, Gang; Liu, Rui; Ooi, Beng Chin; Tan, Kian-Lee, "Blockbench: A Framework for Analyzing Private blockchains," https://arxiv.org/pdf/1703.04057.pdf, March 12, 2017.

CHAPTER 2

How Blockchain Works

We stand at the edge of a new digital revolution. Blockchain probably is the biggest invention since the Internet itself! It is the most promising technology for the next generation of Internet interaction systems and has received extensive attention from many industry sectors as well as academia. Today, many organizations have already realized that they needed to be blockchain ready to sustain their positions in the market. We already looked at a few use cases in Chapter 1, but the possibilities are limitless. Though blockchain is not a silver bullet for all business problems, it has started to impact most business functions and their technology implementations.

To be able to solve some real-world business problems using blockchain, we actually need a fine-grained understanding of what it is and how it works. For this, it needs to be understood through different perspectives such as business, technical, and legal viewpoints. This chapter is an effort to get into the nuts and bolts of blockchain technology and get a complete understanding of how it works.

B. Singhal et al., *Beginning Blockchain*, https://doi.org/10.1007/978-1-4842-3444-0_2

Laying the Blockchain Foundation

Blockchain is not just a technology, it is mostly coupled with business functions and use cases. In its cryptocurrency implementations, it is also interwoven with economic principles. In this section, we will mainly focus on its technical aspects. Technically, blockchain is a brilliant amalgamation of the concepts from cryptography, game theory, and computer science engineering, as shown in Figure 2-1.

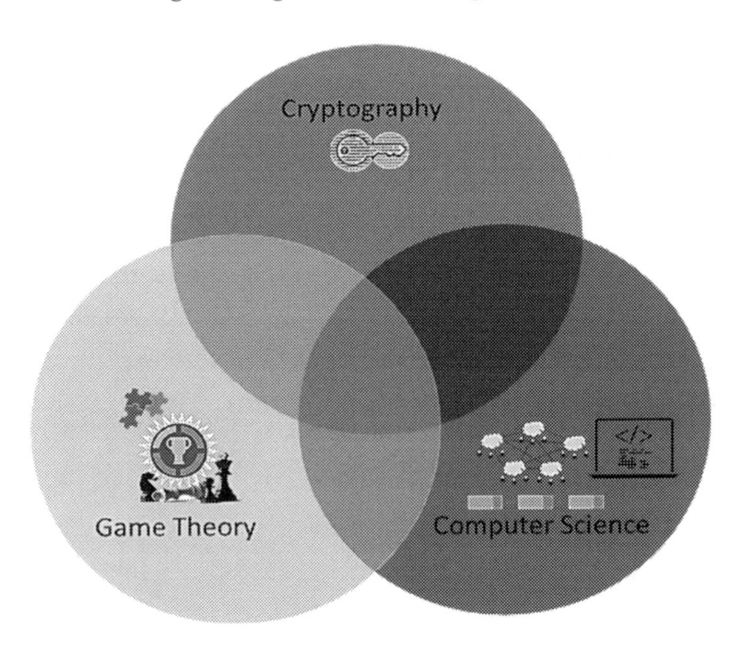

Figure 2-1. *Blockchain at its core*

Let us take a look at what role these components play in the blockchain system at a high level and dig deeper into the fundamentals eventually. Before that, let us quickly revisit how the traditional centralized systems worked. The traditional approach was that there would be a centralized entity that would maintain just one transaction/modification history. This was to exercise concurrency control over the entire database and inject trust into the system through intermediaries. What was the problem with such a stable system then? A centralized system has to be trusted, whether those involved are honest or not! Also, cost due to intermediaries and the transaction time could be greater for obvious reasons. Now think about the centralization of power; having full control of the entire system enables the centralized authorities to do almost anything they want.

Now, let us look at how blockchain addresses these issues due to centralized intermediaries by using cryptography, game theory, and computer science concepts. Irrespective of the use case, the transactions are secured using cryptography. Using cryptography, it can be ensured that a valid user is initiating the transaction and no one can forge a fraudulent transaction. This means, cryptographically it can be ensured that Alice in no way can make a transaction on behalf of Bob by forging his signature. Now, what if a node or a user tries to launch a double-spend attack (e.g., one has just ten bucks and tries to pay the same to multiple people)? Pay close attention here–despite not having sufficient funds, one can still initiate a double-spend, which is cryptographically correct. The only way to prevent double-spend is for every node to be aware of all the transactions. Now this leads to another interesting problem. Since every node should maintain the transaction database, how can they all agree on a common database state? Again, how can the system stay immune to situations where one or more computing nodes deliberately attempt to subvert the system and try to inject a fraudulent database state? The majority of such problems come under the umbrella of the Byzantine Generals' Problem (described later). Well, it gained even more popularity because of blockchain, but it has been there for ages. If you look at the data center solutions, or distributed database solutions, the Byzantine Generals' Problem is an obvious and common problem that they deal with to remain fault tolerant. Such situations and their solution actually come from game theory. The field of game theory provides a radically different approach to determine how a system will behave. The techniques in game theory are arguably the most sophisticated and realistic ones. They usually never consider if a node is honest, malicious, ethical, or has any other such characteristics and believe that the participants act according to the advantage they get, not by moral values. The sole purpose of game theory in blockchain is to ensure that the system is stable (i.e., in Nash Equilibrium) with consensus among the participants.

There are different kinds of business problems and situations with varying degrees of complexities. So, the underlying crypto and game theoretic consensus protocols could be different in different use cases. However, the general principle of maintaining a consistent log or database of verified transactions is the same. Though the concepts of cryptography and game theory have been around for quite some time now, it is the computer science piece that stitches these bits and pieces together through data structures and peer-to-peer network communication technique. Obviously, it is the "smart software engineering" that is needed to realize any logical or mathematical concepts in the digital world. It is then the computer science engineering techniques that incorporate cryptography and game theoretic concepts into an application, enabling decentralized and distributed computing among the nodes with data structure and network communication components.

Cryptography

Cryptography is the most important component of blockchain. It is certainly a research field in itself and is based on advanced mathematical techniques that are quite complex to understand. We will try to develop a solid understanding of some of the cryptographic concepts in this section, because different problems may require different cryptographic solutions; one size never fits all. You may skip some of the details or refer to them as and when needed, but it is the most important component to ensure security in the system. There have been many hacks reported on wallets and exchanges due to weaker design or poor cryptographic implementations.

Cryptography has been around for more than two thousand years now. It is the science of keeping things confidential using encryption techniques. However, confidentiality is not the only objective. There are various other usages of cryptography as mentioned in the following list, which we will explore later:

- **Confidentiality:** Only the intended or authorized recipient can understand the message. It can also be referred to as privacy or secrecy.
- **Data Integrity:** Data cannot be forged or modified by an adversary intentionally or by unintended/accidental errors. Though data integrity cannot prevent the alteration of data, it can provide a means of detecting whether the data was modified.
- **Authentication:** The authenticity of the sender is assured and verifiable by the receiver.
- **Non-repudiation:** The sender, after sending a message, cannot deny later that they sent the message. This means that an entity (a person or a system) cannot refuse the ownership of a previous commitment or an action.

Any information in the form of a text message, numeric data, or a computer program can be called plaintext. The idea is to encrypt the plaintext using an encryption algorithm and a key that produces the ciphertext. The ciphertext can then be transmitted to the intended recipient, who decrypts it using the decryption algorithm and the key to get the plaintext.

Let us take an example. Alice wants to send a message (m) to Bob. If she just sends the message as is, any adversary, say, Eve can easily intercept the message and the confidentiality gets compromised. So, Alice wants to encrypt the message using an encryption algorithm (E) and a secret key (k) to produce the encrypted message called "ciphertext." An adversary has to be aware of both the algorithm (E) and key (k) to intercept the message. The stronger the algorithm and the key, the more difficult it is for the adversary to attack. Note that it would always be desirable to design blockchain systems that are at least provably secure. What this means is that a system must resist certain types of feasible attacks by adversaries.

The common set of steps for this approach can be represented as shown in Figure 2-2.

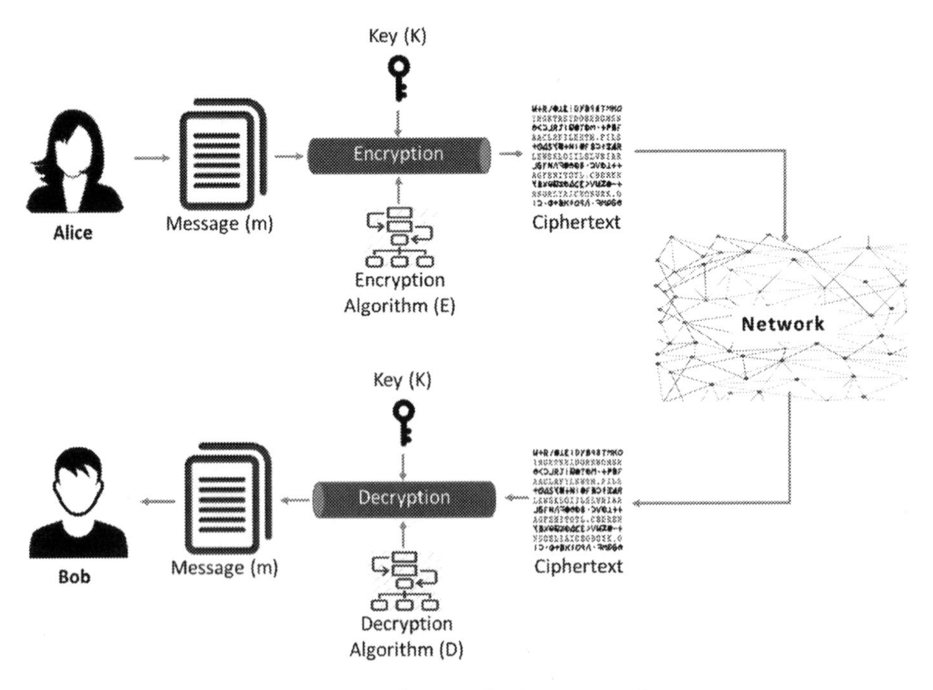

***Figure 2-2.** How Cryptography works in general*

Broadly, there are two kinds of cryptography: symmetric key and asymmetric key (a.k.a. public key) cryptography. Let us look into these individually in the following sections.

Symmetric Key Cryptography

In the previous section we looked at how Alice can encrypt a message and send the ciphertext to Bob. Bob can then decrypt the ciphertext to get the original message. If the same key is used for both encryption and decryption, it is called symmetric key cryptography. This means that both Alice and Bob have to agree on a key (k) called "shared secret" before they exchange the ciphertext. So, the process is as follows:

Alice—the Sender:

- Encrypt the plaintext message m using encryption algorithm E and key k to prepare the ciphertext c
- $c = \text{E}(k, m)$
- Send the ciphertext c to Bob

Bob—the Receiver:

- Decrypt the ciphertext c using decryption algorithm D and the same key k to get the plaintext m
- $m = \text{D}(k, c)$

Did you just notice that the sender and receiver used the same key (k)? How do they agree on the same key and share it with each other? Obviously, they need a secure distribution channel to share the key. It typically looks as shown in Figure 2-3.

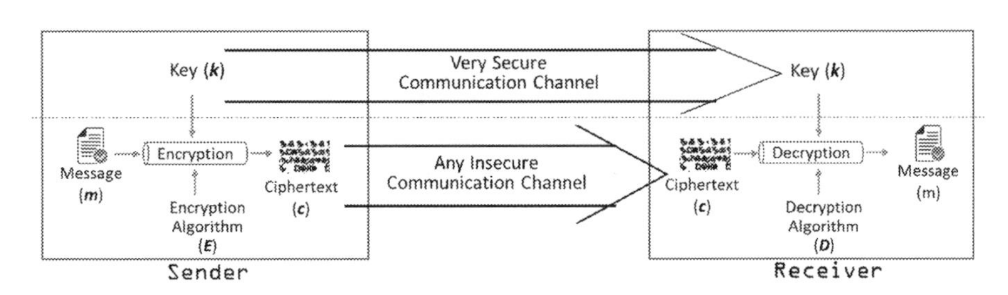

Figure 2-3. *Symmetric cryptography*

Symmetric key cryptography is used widely; the most common uses are secure file transfer protocols such as HTTPS, SFTP, and WebDAVS. Symmetric cryptosystems are usually faster and more useful when the data size is huge.

Please note that symmetric key cryptography exists in two variants: stream ciphers and block ciphers. We will discuss these in the following sections but we will look at Kerchoff's principle and XOR function before that to be able to understand how the cryptosystems really work.

Kerckhoff's Principle and XOR Function

Kerckhoff's principle states that a cryptosystem should be secured even if everything about the system is publicly known, except the key. Also, the general assumption is that the message transmission channel is never secure, and messages could easily be intercepted during transmission. This means that even if the encryption algorithm **E** and decryption algorithm **D** are public, and there is a chance that the message could be intercepted during transmission, the message is still secure due to a shared secret. So, the keys must be kept secret in a symmetric cryptosystem.

The XOR function is the basic building block for many encryption and decryption algorithms. Let us take a look at it to understand how it enables cryptography. The XOR, otherwise known as "Exclusive OR" and denoted by the symbol ⊕, can be represented by the following truth table (Table 2-1).

Table 2-1. *XOR Truth Table*

A	B	A ⊕ B
0	0	0
1	0	1
0	1	1
1	1	0

The XOR function has the following properties, which are important to understand the math behind cryptography:

- **Associative**: $A \oplus (B \oplus C) = (A \oplus B) \oplus C$
- **Commutative**: $A \oplus B = B \oplus A$
- **Negation**: $A \oplus 1 = \bar{A}$
- **Identity**: $A \oplus A = 0$

Using these properties, it would now make sense how to compute the ciphertext "*c*" using plaintext "*m*" and the key "*k*," and then decrypt the ciphertext "*c*" with the same key "*k*" to get the plaintext "*m*." The same XOR function is used for both encryption and decryption.

$m \oplus k = c$ and $c \oplus k = m$

The previous example is in its simplest form to get the hang of encryption and decryption. Notice that it is very simple to get the original plaintext message just by XORing with the key, which is a shared secret and only known by the intended parties. Everyone may know that the encryption or decryption algorithm here is XOR, but not the key.

Stream Ciphers vs. Block Cipher

Stream cipher and block cipher algorithms differ in the way the plaintext is encoded and decoded.

Stream ciphers convert one symbol of plaintext into one symbol of ciphertext. This means that the encryption is carried out one bit or byte of plaintext at a time. In a bit by bit encryption scenario, to encrypt every bit of plaintext, a different key is generated and used. So, it uses an infinite stream of pseudorandom bits as the key and performs the XOR operation with input bits of plaintext to generate ciphertext. For such a system to remain secure, the pseudorandom keystream generator has to be secure and unpredictable. Stream ciphers are an approximation of a proven perfect cipher called "the one-time pad," which we will discuss in a little while.

How does the pseudorandom keystream get generated in the first place? They are typically generated serially from a random seed value using digital shift registers. Stream ciphers are quite simple and faster in execution. One can generate pseudorandom bits offine and decrypt very quickly, but it requires synchronization in most cases.

We saw that the pseudorandom number generator that generates the key stream is the central piece here that ensures the quality of security–which stands to be its biggest disadvantage. The pseudorandom number generator has been attacked many times in the past, which led to deprecation of stream ciphers. The most widely used stream cipher is RC4 (Rivest Cipher 4) for various protocols such as SSL, TLS, and Wi-Fi WEP/WPA etc. It was revealed that there were vulnerabilities in RC4, and it was recommended by Mozilla and Microsoft not to use it where possible.

Another disadvantage is that all information in one bit of input text is contained in its corresponding one bit of ciphertext, which is a problem of low diffusion. It could have been more secured if the information of one bit was distributed across many bits in the ciphertext output, which is the case with block ciphers. Examples of stream ciphers are one-time pad, RC4, FISH, SNOW, SEAL, A5/1, etc.

Block cipher on the other hand is based on the idea of partitioning the plaintext into relatively larger blocks of fixed-length groups of bits, and further encoding each of the blocks separately using the same key. It is a deterministic algorithm with an unvarying transformation using the symmetric key. This means when you encrypt the same plaintext block with the same key, you'll get the same result.

The usual sizes of each block are 64 bits, 128 bits, and 256 bits called block length, and their resulting ciphertext blocks are also of the same block length. We select, say, an $\boldsymbol{r}$ bits key $\boldsymbol{k}$ to encrypt every block of length $\boldsymbol{n}$, then notice here that we have restricted the permutations of the key $\boldsymbol{k}$ to a very small subset of 2^r. This means that the notion of "perfect cipher" does not apply. Still, random selection of the ***r*** bits secret key is important, in the sense that more randomness implies more secrecy.

To encrypt or decrypt a message in block cipher cryptography, we have to put them into a "mode of operation" that defines how to apply a cipher's single-block operation repeatedly to transform amounts of data larger than a block. Well, the mode of operation is not just to divide the data into fixed sized blocks, it has a bigger purpose. We learned that the block cipher is a deterministic algorithm. This means that the blocks with the same data, when encrypted using the same key, will produce the same ciphertext–quite dangerous! It leaks a lot of information. The idea here is to mix the plaintext blocks with the just created ciphertext blocks in some way so that for the same input blocks, their corresponding Ciphertext outputs are different. This will become clearer when we get to the DES and AES algorithms in the following sections.

Note that different modes of operations result in different properties being achieved that add to the security of the underlying block cipher. Though we will not get into the nitty-gritty of modes of operations, here are the names of a few for your reference: Electronic Codebook (ECB), Cipher Block Chaining (CBC), Cipher Feedback (CFB), Output Feedback (OFB), and Counter (CTR).

Block ciphers are a little slow to encrypt or decrypt, compared with the stream ciphers. Unlike stream ciphers where error propagation is much less, here the error in one bit could corrupt the whole block. On the contrary, block ciphers have the advantage of high diffusion, which means that every input plaintext bit is diffused across several ciphertext symbols. Examples of block ciphers are DES, 3DES, AES, etc.

Note A deterministic algorithm is an algorithm that, given a particular input, will always produce the same output.

One-Time Pad

It is a symmetric stream cipher where the plaintext, the key, and the ciphertext are all bit strings. Also, it is completely based on the assumption of a "purely random" key (and not pseudorandom), using which it could achieve "perfect secrecy." Also, as per the design, the key can be used only once. The problem with this is that the key should be at least as long as the plaintext. It means that if you are encrypting a 1GB file, the key would also be 1GB! This gets impractical in many real-world cases.

Example:

Table 2-2. *Example Encryption Using XOR Function*

PlainText	**1**	**0**	**0**	**1**	**1**	**1**	**0**	**0**	**1**	**0**	**1**	**0**	**1**	**1**	**0**	**1**	**1**	**0**
Key	0	1	0	0	1	1	0	1	1	1	0	0	1	0	1	0	1	1
Ciphertext	1	1	0	1	0	0	0	1	0	1	1	0	0	1	1	1	0	1

You can refer to the XOR truth table in the previous section to find how ciphertext is generated by XOR-ing plaintext with the key. Notice that the plaintext, the key, and the ciphertext are all 18 bits long.

Here, the receiver upon receipt of the ciphertext can simply XOR again with the key and get the plaintext. You can try it on your own with Table 2-2 and you will get the same plaintext.

The main problem with this one-time pad is more of practicality, rather than theory. How do the sender and receiver agree on a secret key that they can use? If the sender and the receiver already have a secure channel, why do they even need a key? If they do not have a secure channel (that is why we use cryptography), then how can they share the key securely? This is called the "key distribution problem."

The solution is to approximate the one-time pad by using a pseudorandom number generator (PRNG). This is a deterministic algorithm that uses a seed value to generate a sequence of random numbers that are not truly random; this in itself is an issue. The sender and the receiver have to have the same seed value for this system to work. Sharing that seed value is way better compared with sharing the entire key; just that it has to be secured. It is susceptible to compromise by someone who knows the algorithm as well as the seed.

Data Encryption Standard

The Data Encryption Standard (DES) is a symmetric block cipher technique. It uses 64-bit block size with a 64-bit key for encryption and decryption. Out of the 64-bit key, 8 bits are reserved for parity checks and technically 56 bits is the key length. It has been proven that it is vulnerable to brute force attack and could be broken in less than a day. Given Moore's law, it could be broken a lot quicker in the future, so its usage has been deprecated quite a bit because of the key length. It was very popular and was being used in banking applications, ATMs, and other commercial applications, and more so in hardware implementations than software. We give a high-level description of the DES encryption in this section.

In symmetric cryptography, a large number of block ciphers use a design scheme known as a "Feistel cipher" or "Feistel network." A Feistel cipher consists of multiple rounds to process the plaintext with the key, and every round consists of a substitution step followed by a permutation step. The more the number of rounds, the more secure it could be but encryption/ decryption gets slower. The DES is based on a Feistel cipher with 16 rounds. A general sequence of steps in the DES algorithm is shown in Figure 2-4.

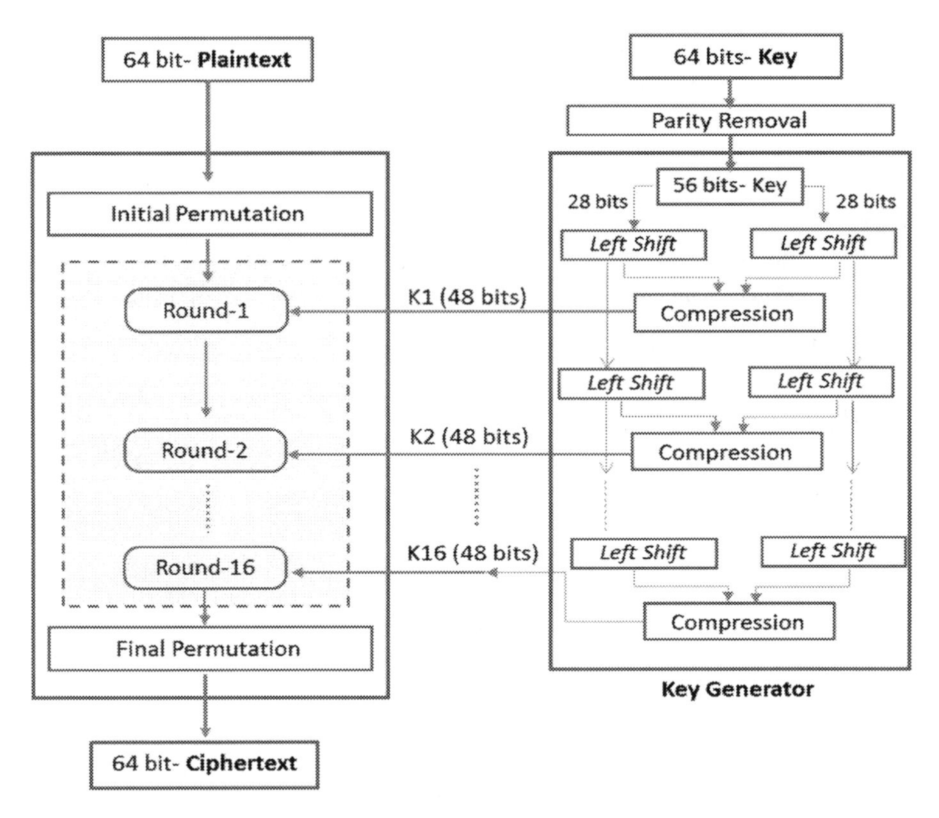

Figure 2-4. *DES cryptography*

Let us first talk about the key generator and then we will get into the encryption part.

- As mentioned before, the key is also 64 bits long. Since 8 bits are used as parity bits (more precisely, bit number 8, 16, 24, 32, 40, 48, 56, and 64), only 56 bits are used for encryption and decryption.
- After parity removal, the 56-bit key is divided into two blocks, each of 28 bits. They are then bit-wise left shifted in every round. We know that the DES uses 16 rounds of Feistel network. Note here that every round

takes the previous round's left-shifted bit block and then again left shifts by one bit in the current round.

- Both the left-shifted 28-bit blocks are then combined through a compression mechanism that outputs a 48-bit key called subkey that gets used for encryption. Similarly, in every round, the two 28-bit blocks from the previous round get left shifted again by one bit and then clubbed and compressed to the 48-bit key. This key is then fed to the encryption function of the same round.

Let us now look at how DES uses the Feistel cipher rounds for encryption:

- First, the plaintext input is divided into 64 bit blocks. If the number of bits in the message is not evenly divisible by 64, then the last block is padded to make it a 64-bit block.
- Every 64-bit input data block goes through an initial permutation (IP) round. It simply permutes, i.e., rearranges all the 64-bit inputs in a specific pattern by transposing the input blocks. It has no cryptographic significance as such, and its objective is to make it easier to load plaintext/ciphertext into DES chips in byte-sized format.
- After the IP round, the 64-bit block gets divided into two 32-bit blocks, a left block (L) and a right block (R). In every round, the blocks are represented as L_i and R_i, where the subscript "I" denotes the round. So, the outcomes of IP round are denoted as L_0 and R_0.

- Now the Feistel rounds start. The first round takes L_0 and R_0 as input and follows the following steps:
  - The right side 32-bit block (R) comes as is to the left side and the left side 32-bit block (L) goes through an operation with the key ***k*** of that round and the right side 32-bit block (R) as shown following:
  - $L_i = R_i - 1$
  - $R_i = L_i - 1 \oplus F(R_i - 1, K_i)$ where "I" is the round number
  - The F() is called the "Cipher Function" that is actually the core part of every round. There are multiple steps or operations that are bundled together in this F() operation.
  - In the first step, operation of the 32-bit R-block is expanded and permuted to output a 48-bit block.
  - In the second step, this 48-bit block is then XORed with the 48-bit subkey supplied by the key generator of the same round.
  - In the third step, this 48-bit XORed output is fed to the substitution box to reduce the bits back to 32 bits. The substitution operation in this S-box is the only nonlinear operation in DES and contributes significantly to the security of this algorithm.
  - In the fourth step, the 32-bit output of the S-box is fed to the permutation box (P-box), which is just a permutation operation that outputs a 32-bit block, which is actually the final output of F() cipher function.
  - The output of F() is then XORed with the 32-bit L-block, which is input to this round. This XORed output then becomes the final R-block output of this round.

- Refer to Figure 2-5 to understand the various operations that take place in every round.

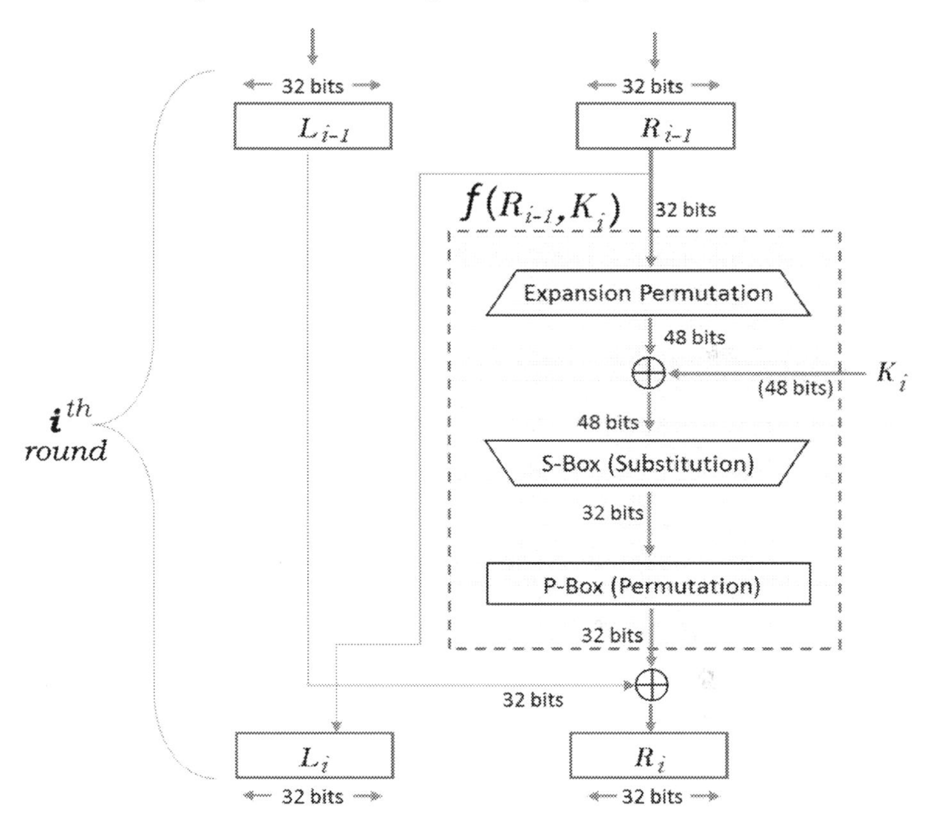

Figure 2-5. *Round function of DES*

- The previously discussed Feistel round gets repeated 16 times, where the output of one round is fed as the input to the following round.
- Once all the 16 rounds are over, the output of the 16th round is again swapped such that the left becomes the right block and vice versa.
- Then the two blocks are clubbed to make a 64-bit block and passed through a permutation operation, which is the inverse of the initial permutation function and that results in the 64-bit ciphertext output.

We looked at how the DES algorithm really works. The decryption also works a similar way in the reverse order. We will not get into those details, but leave it to you to explore.

Let us conclude with the limitations of the DES. The 56-bit key length was susceptible to brute force attack and the S-boxes used for substitution in each round were also prone to cryptanalysis attack because of some inherent weaknesses. Because of these reasons, the Advanced Encryption Standard (AES) has replaced the DES to the extent possible. Many applications now choose AES over DES.

Advanced Encryption Standard

Like DES, the AES algorithm is also a symmetric block cipher but is not based on a Feistel network. The AES uses a substitution-permutation network in a more general sense. It not only offers greater security, but also offers greater speed! As per the AES standards, the block size is fixed at 128 bits and allows a choice of three keys: 128 bits, 192 bits, and 256 bits. Depending on the choice of the key, AES is named as AES-128, AES-192, and AES-256.

In AES, the number of encryption rounds depend on the key length. For AES-128, there are ten rounds; for AES-192, there are 12 rounds; and for AES-256, there are 14 rounds. In this section, our discussion is limited to key length 128 (i.e., AES-128), as the process is almost the same for other variants of AES. The only thing that changes is the "key schedule," which we will look into later in this section.

Unlike DES, AES encryption rounds are iterative and operate an entire data block of 128 bits in every round. Also, unlike DES, the decryption is not very similar to the encryption process in AES.

To understand the processing steps in every round, consider the 128-bit block as 16 bytes where individual bytes are arranged in a 4×4 matrix as shown:

Byte 0	*Byte 4*	*Byte 8*	*Byte 12*
Byte 1	*Byte 5*	*Byte 9*	*Byte 13*
Byte 2	*Byte 6*	*Byte 10*	*Byte 14*
Byte 3	*Byte 7*	*Byte 11*	*Byte 15*

This 4 × 4 matrix of bytes as shown is referred to as **state array**. Please note that every round consumes an input state array and produces an output state array.

The AES also uses another piece of jargon called "word" that needs to be defined before we go further. Whereas a byte consists of eight bits, a word consists of four bytes, that is, 32 bits. The four bytes in each column of the state array form 32-bit words and can be called **state words**. The state array can be shown as follows:

$word_0$	$word_1$	$word_2$	$word_3$
Byte 0	*Byte 4*	*Byte 8*	*Byte 12*
Byte 1	*Byte 5*	*Byte 9*	*Byte 13*
Byte 2	*Byte 6*	*Byte 10*	*Byte 14*
Byte 3	*Byte 7*	*Byte 11*	*Byte 15*

Also, every byte can be represented with two hexadecimal numbers. Example: if the 8-bit byte is {00111010}, it could be represented as "3A" in Hex notation. "3" represents the left four bits "0011" and "A" represents the right four bits "1010."

Now to generalize each round, processing in each round happens at byte level and consists of byte-level substitution followed by word-level permutation, hence it is a substitution-permutation network. We will get to further details when we discuss the various operations in each round. The overall encryption and decryption process of AES can be represented in Figure 2-6.

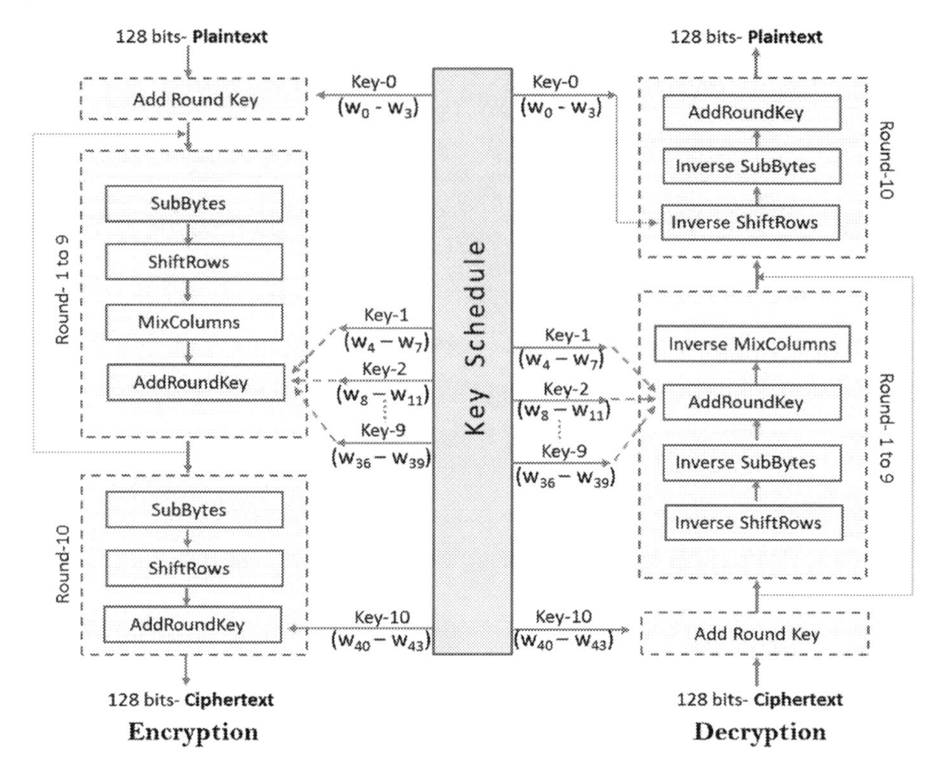

Figure 2-6. *AES cryptography*

If you paid close attention to Figure 2-6, you would have noticed that the decryption process is not just the opposite of encryption. The operations in the rounds are executed in a different order! All steps of the round function–SubBytes, ShiftRows, MixColumns, AddRoundKey–are invertible. Also, notice that the rounds are iterative in nature. Round 1 through round 9 have all four operations, and the last round excludes only the "MixColumns" operation. Let us now build a high-level understanding of each operation that takes place in a round function.

SubBytes: This is a substitution step. Here, each byte is represented as two hexadecimal digits. As an example, take a byte {00111010} represented with two hexadecimal digits, say {3A}. To find its substitution values, refer to the S-box lookup table (16 × 16 table) to find the corresponding value for 3-row and A-column. So, for {3A}, the corresponding substituted value would be {80}. This step provides the nonlinearity in the cipher.

ShiftRows: This is the transformation step and is based upon the matrix representation of the state array. It consists of the following shift operations:

- No circular shifting of the first row, and remains as is
- Circularly shifting of the second row by one byte to the left
- Circularly shifting of the third row by two bytes to the left
- Circularly shifting of the fourth row (last row) by three bytes to the left

It can be represented as shown:

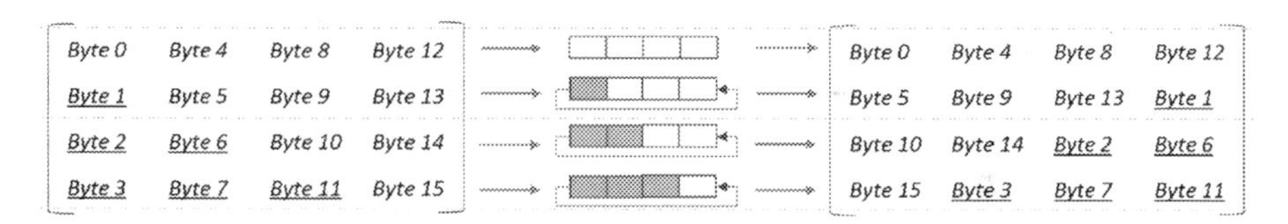

MixColumns: It is also a transformation step where all the four columns of the state are multiplied with a fixed polynomial ($\boldsymbol{C_x}$) and get transformed to new columns. In this process, each byte of a column is mapped to a new value that is a function of all four bytes in the column. This is achieved by the matrix multiplication of state as shown:

$$\begin{bmatrix} 2 & 3 & 1 & 1 \\ 1 & 2 & 3 & 1 \\ 1 & 1 & 2 & 3 \\ 3 & 1 & 1 & 2 \end{bmatrix} \begin{bmatrix} \text{Byte } 0 & \text{Byte } 4 & \text{Byte } 8 & \text{Byte } 12 \\ \text{Byte } 1 & \text{Byte } 5 & \text{Byte } 9 & \text{Byte } 13 \\ \text{Byte } 2 & \text{Byte } 6 & \text{Byte } 10 & \text{Byte } 14 \\ \text{Byte } 3 & \text{Byte } 7 & \text{Byte } 11 & \text{Byte } 15 \end{bmatrix} = \begin{bmatrix} \text{Byte } 0' & \text{Byte } 4' & \text{Byte } 8' & \text{Byte } 12' \\ \text{Byte } 1' & \text{Byte } 5' & \text{Byte } 9' & \text{Byte } 13' \\ \text{Byte } 2' & \text{Byte } 6' & \text{Byte } 10' & \text{Byte } 14' \\ \text{Byte } 3' & \text{Byte } 7' & \text{Byte } 11' & \text{Byte } 15' \end{bmatrix}$$

The matrix multiplication is as usual, but the AND products are XORed. Let us see one of the examples to understand the process. Byte 0′ is calculated as shown:

Byte 0′ = (2 . Byte0) ⊕ (3 . Byte1) ⊕ Byte3 ⊕ Byte4

It is important to note that this MixColumns step, along with the ShiftRows step, provide the necessary diffusion property (information from one byte gets diffused to multiple bytes) to the cipher.

AddRoundKey: This is again a transformation step where the 128-bit round key is bitwise XORed with 128 bits of state in a column major order. So, the operation takes place column-wise, meaning four bytes of a word state column with one word of the round key. In the same way we represented the 128-bit plaintext block, the 128-bit key should also be represented in the same 4 × 4 matrix as shown here:

w_0	w_1	w_2	w_3
Byte 0	Byte 4	Byte 8	Byte 12
Byte 1	Byte 5	Byte 9	Byte 13
Byte 2	Byte 6	Byte 10	Byte 14
Byte 3	Byte 7	Byte 11	Byte 15

128-bit key

This operation affects every bit of a state. Now, recollect that there are ten rounds, and each round has its own round key. Since there is an "AddRoundKey" step before the rounds start, effectively there are eleven (10 + 1) AddRoundKey operations. In one round, all 128-bits of subkey, that is, all four words of subkey, are used to XOR with the 128-bit input data block. So, in total, we require 44 key words, $\mathbf{w}_0$ through $\mathbf{w}_{43}$. This is why the 128-bit key has to go through a key expansion operation, which we will get to in a little while.

Note here that the key word [$\mathbf{w}_0$, $\mathbf{w}_1$, $\mathbf{w}_2$, $\mathbf{w}_3$] gets XORed with the initial input block before the round-based processing begins. The remaining 40 word-keys, $\mathbf{w}_4$ through $\mathbf{w}_{43}$, get used four words at a time in each of the ten rounds.

AES Key Expansion: The AES key expansion algorithm takes as input a 128-bit cipher key (four-word key) and produces a schedule of 44 key words from it. The idea is to design this system in such a way that a one-bit change in the key would significantly affect all the round keys.

The key expansion operation is designed such that each grouping of a four-word key produces the next grouping of a four-word key in a four-word to four-word basis. It is easy to explain this with a pictorial representation, so here we go:

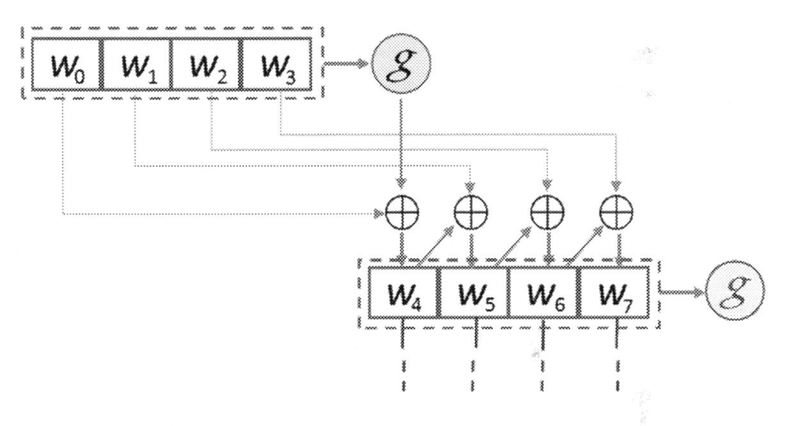

***Figure 2-7.** AES key expansion*

We will quickly run through the operations that take place for key expansion by referring to the diagram:

- The initial 128-bit key is [$\mathbf{w_0}$, $\mathbf{w_1}$, $\mathbf{w_2}$, $\mathbf{w_3}$] arranged in four words.
- Take a look at the expanded word now: [$\mathbf{w_4}$, $\mathbf{w_5}$, $\mathbf{w_6}$, $\mathbf{w_7}$]. Notice that $\mathbf{w_5}$ depends on $\mathbf{w_4}$ and $\mathbf{w_1}$. This means that every expanded word depends on the immediately preceding word, i.e., $\mathbf{w_i}$ – 1 and the word that is four positions back, i.e., $\mathbf{w_i}$ – 4 . Test the same for $\mathbf{w_6}$ as well. As you can see, just a simple XOR operation is performed here.

- Now, what about $\mathbf{w_4}$? Or, any other position that is a multiple of four, such as $\mathbf{w_8}$ or $\mathbf{w_{12}}$? For these words, a more complex function denoted as "**g**" is used. It is basically a three-step function. In the first step, the input four-word block goes through circular left shift by one byte. For example [$\mathbf{w_0}$, $\mathbf{w_1}$, $\mathbf{w_2}$, $\mathbf{w_3}$] becomes [$\mathbf{w_1}$, $\mathbf{w_2}$, $\mathbf{w_3}$, $\mathbf{w_0}$]. In the second step, the four bytes input word (e.g., [$\mathbf{w_1}$, $\mathbf{w_2}$, $\mathbf{w_3}$, $\mathbf{w_0}$]) is taken as input and byte substitution is applied on each byte using S-box. Then, in the third step, the result of step 2 is XORed with something called round constant denoted as Rcon[ ]. The round constant is a word in which the right-most three bytes are always zero. For example, [x, 0, 0, 0]. This means that the purpose of Rcon[ ] is to just perform XOR on the left-most byte of the step 2 output key word. Also note that the Rcon[ ] is different for each round. This way, the final output of the complex function "**g**" is generated, which is then XORed with $\mathbf{w_i}$ - 4 to get $\mathbf{w_i}$ where "I" is a multiple of 4.
- This is how the key expansion takes place in AES.

The output state array of the last round is rearranged back to form the 128-bit ciphertext block. Similarly, the decryption process takes place in a different order, which we looked at in the AES process diagram. The idea was to give you a heads-up on how this algorithm works at a high level, and we will restrict our discussion to just the encryption process in this section.

The AES algorithm is standardized by the NIST (National Institute of Standards and Technology). It had the limitation of long processing time. Assume that you are sending just a 1 megabyte file (8388608 bits) by encrypting with AES. Using a 128-bit AES algorithm, the number of steps required for this encryption will be 8388608/128 = 65536 on this same number of data blocks! Using a parallel processing approach, AES efficiency can be increased, but is still not very suitable when you are dealing with large data.

Challenges in Symmetric Key Cryptography

There are some limitations in symmetric key cryptography. A few of them are listed as follows:

- The key must be shared by the sender and receiver before any communication. It requires a secured key establishment mechanism in place.
- The sender and receiver must trust each other, as they use the same symmetric key. If a receiver is hacked by an attacker or the receiver deliberately shared the key with someone else, the system gets compromised.
- A large network of, say, n nodes require key $n(n-1)/2$ key pairs to be managed.
- It is advisable to keep changing the key for each communication session.
- Often a trusted third party is needed for effective key management, which itself is a big issue.

Cryptographic Hash Functions

Hash functions are the mathematical functions that are the most important cryptographic primitives and are an integral part of blockchain data structure. They are widely used in many cryptographic protocols, information security applications such as Digital Signatures and message authentication codes (MACs). Since it is used in asymmetric key cryptography, we will discuss it here prior to getting into asymmetric cryptography. Please note that the concepts covered in this section may not be in accordance with the academic text books, and a little biased toward the blockchain ecosystem.

Cryptographic hash functions are a special class of hash functions that are apt for cryptography, and we will limit our discussion pertaining to it only. So, a cryptographic hash function is a one-way function that converts input data of arbitrary length and produces a fixed-length output. The output is usually termed "hash value" or "message digest." It can be represented as shown Figure 2-8.

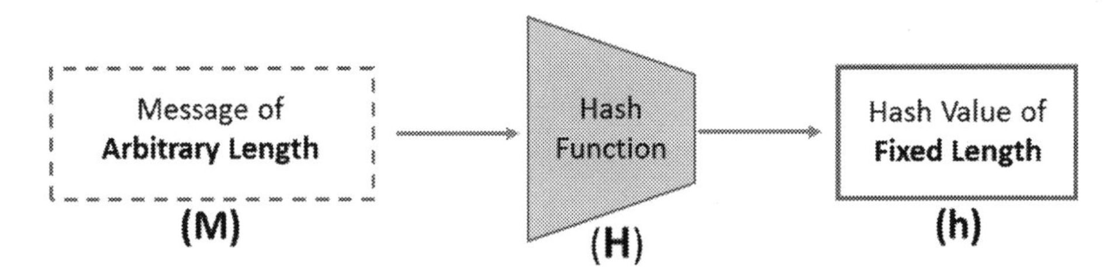

Figure 2-8. *Hash function in its basic form*

For the hash functions to serve their design purpose and be usable, they should have the following core properties:

- Input can be any string of any size, but the output is of fixed length, say, a 256-bit output or a 512-bit output as examples.
- The hash value should be efficiently computable for any given message.
- It is deterministic, in the sense that the same input when provided to the same hash function produces the same hash value every time.
- It is infeasible (though not impossible!) to invert and generate the message from its hash value, except trying for all possible messages.

- Any small change in the message should greatly influence the output hash, just so no one can correlate the new hash value with the old one after a small change.

Apart from the aforementioned core properties, they should also meet the following security properties to be considered as a cryptographic protocol:

- Collision resistance: It implies that it is infeasible to find two different inputs, say, X and Y, that hash to the same value.

$$X \longrightarrow H \longrightarrow H(X), \quad Y \longrightarrow H \longrightarrow H(Y) \qquad H(X) \neq H(Y), \qquad \text{Given } (X \neq Y)$$

This makes the hash function H() collision resistant because no one can find X and Y, such that H(X) = H(Y). Note that this hash function is a compression function, as it compresses a given input to fixed sized output that is shorter than the input. So, the input space is too large (anything of any size) compared with the output space, which is fixed. If the output is a 256-bit hash value, then the output space can have a maximum of 2^{256} values, and not beyond that. This implies that a collision must exist. However, it is extremely difficult to find that collision. As per the theory of "the birthday paradox," we can infer that it should be possible to find a collision by using the square root of the output space. So, by taking $2^{128} + 1$ inputs, it is highly likely to find a collision; but that is an extremely huge number to compute, which is quite infeasible!

Let us now discuss where this property could be useful. In the majority of online storage, cloud file storage, blob storage, App Stores, etc., the property "collision resistance" is widely used to ensure the integrity of the files. Example: someone computes the message digest of a file and uploads to cloud storage. Later when they download the file, they could just compute the message digest again and cross-check with the old one they have. This way, it can be ensured if the file was corrupted because of some transmission issues or possibly due to some deliberate attempts. It is due to the property of collision resistance that no one can come up with a different file or a modified file that would hash to the same value as that of the original file.

- Preimage resistance: This property means that it is computationally impossible to invert a hash function; i.e., finding the input X from the output H(X) is infeasible. Therefore, this property can also be called "hiding" property. Pay close attention here; there is another subtle aspect to this situation. Note that when X can be anything in the world, this property is easily achieved. However, if there are just a limited number of values that X can take, and that is known to the adversary, they can easily compute all possible values of X and find which one hashes to the outcome.

  Example: A laboratory decided to prepare the message digests for the successful outcome of an experiment so that any adversary who gets access to the results database cannot make any sense of it because what is stored in the system are hashed outputs. Assume that there can only be three possible outcomes of the experiment such as OP111, OP112, and OP113, out of which only one is successful, say, OP112. So, the laboratory decides to hash it, compute H(OP112), and store the hashed values in the system. Though an adversary cannot find OP112 from H(OP112), they can simply hash all the possible outcomes of the experiment, i.e., H(OP111), H(OP112), and H(OP113) and see that only H(OP112) is matching with what is stored in the system. Such a situation is certainly vulnerable! This means that, when the input to a hash function comes from a limited space and does not come from a spread-out distribution, it is weak. However, there is a solution to it as follows:

Let us take an input, say "**X**" that is not very spread out, just like the outcomes of the experiment we just discussed with a few possible values. If we can concatenate that with another random input, say "**r**," that comes from a probability distribution with high min entropy, then it will be difficult to find X from H(**r** || **X**). Here, high min entropy means that the distribution is very spread out and there is no particular value that is likely to occur. Assume that "**r**" was chosen from 256-bit distribution. For an adversary to get the exact value of "**r**" that was used along with input, there is a success probability of $1/2^{256}$, which is almost impossible to achieve. The only way is to consider all the possible values of this distribution one by one—which is again practically impossible. The value "**r**" is also referred to as "nonce." In cryptography, a nonce is a random number that can be used only once.

Let us now discuss where this property of preimage resistance could be useful. It is very useful in committing to a value, so "commitment" is the use case here. This can be better explained with an example. Assume that you have participated in some sort of betting or gambling event. Say you have to commit to your option, and declare it as well. However, no one should be able to figure out what you are betting on, and you yourself cannot deny later on what you bet on. So, you leverage the preimage resistance property of Hash Function. You take a hash of the choice you are betting on, and declare it publicly. No one can invert the hash function and figure out what you are betting on. Also, you cannot later say that your choice was different, because if you hash a different choice, it will not match what you have declared publicly. It is advisable to use a nonce "**r**" the way we explained in the previous paragraph to design such systems.

- Second preimage resistance: This property is slightly different from "collision resistant." It implies that given an input X and its hash H(X), it is infeasible to find Y, such that H(X) = H(Y). Unlike in collision-resistant property, here the discussion is for a given X, which is fixed. This implies that if a hash function is collision resistant already, then it is second preimage resistant also.

There is another derived property from the properties mentioned that is quite useful in Bitcoin. Let us look into it from a technical point of view and learn how Bitcoin leverages it for mining when we hit Chapter 3. The name of this property is "puzzle friendliness." This name implies that there is no shortcut to the solution and the only way to get to the solution is to traverse through all the possible options in the input space. We will not try to define it here but will directly try to understand what it really means. Let us consider this example: H(**r** || **X**) = **Z**, where "**r**" is chosen from a distribution with high min entropy, "**X**" is the input concatenated with "**r**," and "**Z**" is the hashed output value. The property means that it is way too hard for an adversary to find a value "**Y**" that exactly hashes to "**Z**." That is, H(**r′** || **Y**) = **Z**, where **r′** is a part of the input chosen in the same randomized way as "**r**." What this means is that, when a part of the input is substantially randomized, it is hard to break the hash function with a quick solution; the only way is to test with all possible random values.

In the previous example, if "**Z**" is an **n**-bits output, then it has taken just one value out of 2^n possible values. Note carefully that a part of your input, say "**r**," is from a high min-entropy distribution, which has to be appended with your input **X**. Now comes the interesting part of designing a search puzzle. Let's say **Z** is an **n**-bits output and is a set of 2^n possible values, not just an exact value. You are asked to find a value of **r** such that when hashed appended with **X**, it falls within that output set of 2^n values; then it forms a search puzzle. The idea is to find all possible values of **r** till it falls withing the range of **Z**. Note here that the size of **Z** has limited the output space to a smaller set of 2^n possible values. The smaller the output space, the harder is the problem. Obviously, if the range is big, it is easier to find a value in it and if the range is quite narrow with just a few possibilities, then finding a value within that range is tough. This is the beauty of the "**r**," called the "nonce" in the input to hash function. Whatever random value of **r** you take, it will be concatenated with "**X**" and will go through the same hash function, again and again, till you get the right nonce value "**r**" that satisfies the required range for **Z**, and there are absolutely no shortcuts to it except for trying all possible values!

Note that for an **n**-bit hash value output, an average effort of 2^n is needed to break preimage and second preimage resistance, and $2^n/2$ for collision resistance.

We discussed various fundamental and security properties of hash functions. In the following sections we will see some important hash functions and dive deeper as applicable.

A Heads-up on Different Hash Functions

One of the oldest hash functions or compression function is the MD4 hash function. It belongs to the message digest (MD) family. Other members of the MD family are MD5 and MD6, and there are many other variants of MD4 such as RIPEMD. The MD family of algorithms produce a 128-bit message digest by consuming 512-bit blocks. They were widely used as checksums to verify data integrity. Many file servers or software repositories used to provide a precomputed MD5 checksum, which the users could check against the file they downloaded. However, there were a lot of vulnerabilities found in the MD family and it was deprecated.

Another such hash function family is the Secure Hash Algorithm (SHA) family. There are basically four algorithms in this family, such as SHA-0, SHA-1, SHA-2, and SHA-3. The first algorithm proposed in this family was named SHA, but newer versions were coming with security fixes and updates, so a retronym was applied to it and it was made SHA-0. It was found to have a serious yet undisclosed security flaw and was discontinued. Later, SHA-1 was proposed as a replacement to SHA-0. SHA-1 had an extra computational step that addressed the problem in SHA-0. Both SHA-0 and SHA-1 were 160-bit hash functions that consumed 512-bit block sizes. SHA-1 was designed by the National Security Agency (NSA) to use it in the digital signature algorithm (DSA). It was used quite a lot in many security tools and Internet protocols such as SSL, SSH, TSL, etc. It was also used in version control systems such as Mercurial, Git, etc. for consistency checks, and not really for security. Later, around 2005, cryptographic weaknesses were found in it and it was deprecated after the year 2010. We will get into SHA-2 and SHA-3 in detail in the following sections.

SHA-2

It belongs to the SHA family of hash functions, but itself is a family of hash functions. It has many SHA variants such as SHA-224, SHA-256, SHA-384, SHA-512, SHA-512/224, and SHA-512/256. SHA-256 and SHA-512 are the primitive hash functions and the other variants are derived from them. The SHA-2 family of hash functions are widely used in applications such as SSL, SSH, TSL, PGP, MIME, etc.

SHA-224 is a truncated version of SHA-256 with a different initial value or initialization vector (IV). Note that the SHA variants with different truncations applied can produce the same bit length hash outputs, hence different initialization vectors are applied in different SHA variants to be able to properly differentiate them. Now coming back to the SHA-224 computation, it is a two-step process. First, SHA-256 value is computed with a different IV compared with the default one used in SHA-256. Second, the resulting 256-bit hash value is truncated to 224-bit; usually the 224 bits from left are kept, but the choice is all yours.

SHA-384 is a truncated version of SHA-512, just the way SHA-224 is a truncated version of SHA-256. Similarly, both 512/224 and SHA-512/256 are truncated versions of SHA-512. Are you wondering why this concept of "truncation" exists? Note that truncation is not just limited to the ones we just mentioned, and there can be various other variants as well. The primary reasons for truncation could be as follows:

- Some applications require a message digest with a certain length that is different from the default ones.
- Irrespective of the SHA-2 variant we are using, we can select a truncation level depending on what security property we want to sustain. Example: Considering today's state of computing power, when collision resistance is necessary, we should keep at least 160 bits and when only preimage-resistance is necessary,

we should keep at least 80 bits. The security property such as collision resistance decreases with truncation, but it should be chosen such that it would be computationally infeasible to find a collision.

- Truncation also helps maintain the backward compatibility with older applications. Example: SHA-224 provides 112-bit security that can match the key length of triple-DES (3DES).

Talking about efficiency, SHA-256 is based on a 32-bit word and SHA-512 is based on a 64-bit word. So, on a 64-bit architecture, SHA-512 and all its truncated variants can be computed faster with a better level of security compared with SHA-1 or other SHA-256 variants.

Table 2-3 is a tabular representation taken from the NIST paper that represents SHA-1 and different SHA-2 algorithms properties in a nutshell.

Table 2-3. *SHA-1 & SHA-2 Hash Function in a Nutshell*

Algorithm	Message Size (bits)	Block Size (bits)	Word Size (bits)	Message Digest Size (bits)
SHA-1	$<2^{64}$	512	32	160
SHA-224	$<2^{64}$	512	32	224
SHA-256	$<2^{64}$	512	32	256
SHA-384	$<2^{128}$	1024	64	384
SHA-512	$<2^{128}$	1024	64	512
SHA-512/224	$<2^{128}$	1024	64	224
SHA-512/256	$<2^{128}$	1024	64	256

As a rule of thumb, it is advisable not to truncate when not necessary. Certain hash functions tolerate truncation and some don't, and it also depends on how you are using it and in what context.

SHA-256 and SHA-512

As mentioned already, SHA-256 belongs to the SHA-2 family of hash functions, and this is the one used in Bitcoins! As the name suggests, it produces a 256-bit hash value, hence the name. So, it can provide 2^{128}-bit security as per the birthday paradox.

Recall that the hash functions take arbitrary length input and produce a fixed size output. The arbitrary length input is not fed as is to the compression function and is broken into fixed length blocks before it is fed to the compression function. This means that a construction method is needed that can iterate through the compression function by constructing fixed-sized input blocks from arbitrary length input data and produce a fixed length output. There are various types of construction methods such as Merkle-Damgård construction, tree construction, and sponge construction. It is proven that if the underlying compression function is collision resistant, then the overall hash function with any construction method should also be collision resistant.

The construction method that SHA-256 uses is the Merkle-Damgård construction, so let us see how it works in Figure 2-9.

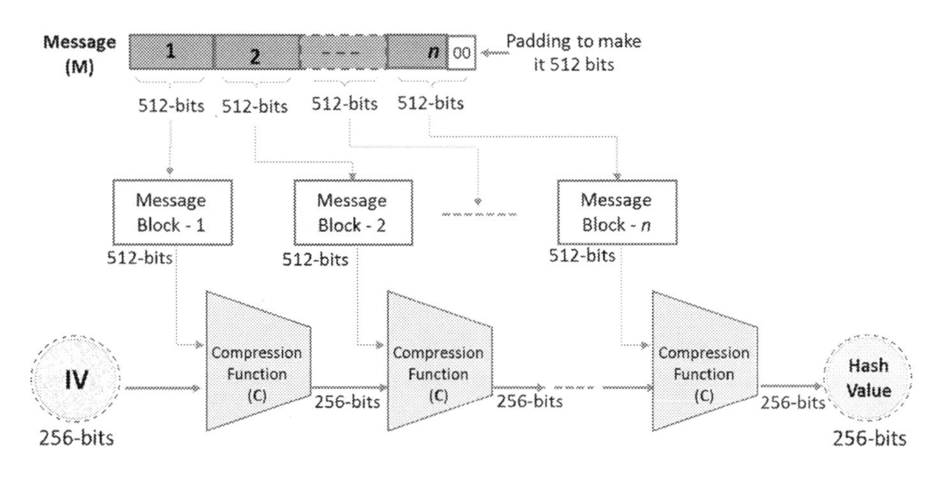

Figure 2-9. *Merkle-Damgård construction for SHA-256*

Referring to the diagram, the following steps (presented at a high level) are executed in the order specified to compute the final hash value:

- As you can see in the diagram, the message is first divided into 512-bit blocks. When the message is not an exact multiple of 512 bits (which is usually the case), the last block falls short of bits, hence it is padded to make it 512 bits.
- The 512-bit blocks are further divided into 16 blocks of 32-bit words ($16 \times 32 = 512$).
- Each block goes through 64 rounds of round function where each 32-bit word goes through a series of operations. The round functions are a combination of some common functions such as XOR, AND, OR, NOT, Bit-wise Left/Right Shift, etc. and we will not get into those details in this book.

Similar to SHA-256, the steps and the operations are quite similar in SHA-512, as SHA-512 also uses Merkle-Damgård construction. The major difference is that there are 80 rounds of round functions in SHA-512 and the word length is 64 bits. The block size in SHA-512 is 1024 bits, which gets further divided into 16 blocks of 64-bit words The output message digest is 512 bits in length, that is, eight blocks of 64-bit words. While SHA-512 was gaining momentum, and started being used in many applications, a few people turned to the SHA-3 algorithm to be future ready. SHA-3 is just a different approach to hashing and not a real replacement to SHA-256 or SHA-512, though it allows tuning. We will learn a few more details about SHA-3 in the following sections.

RIPEMD

RACE Integrity Primitives Evaluation Message Digest (RIPEMD) hash function is a variant of the MD4 hash function with almost the same design considerations. Since it is used in Bitcoins, we will have a brief discussion on it.

The original RIPEMD was of 128 bits, later RIPEMD-160 was developed. There exist 128-, 256-, and 320-bit versions of this algorithm, called RIPEMD-128, RIPEMD-256, and RIPEMD-320, respectively, but we will limit our discussion to the most popular and widely used RIPEMD-160.

RIPEMD-160 is a cryptographic hash function whose compression function is based on the Merkle–Damgård construction. The input is broken into 512-bit blocks and padding is applied when the input bits are not a multiple of 512. The 160-bit hash value output is usually represented as 40-digit hexadecimal numbers.

The compression function is made up of 80 stages, made up of two parallel lines of five rounds of 16 steps each (5 × 16 = 80). The compression function works on sixteen 32-bit words (512-bit blocks).

Note Bitcoin uses both SHA-256 and RIPEMD-160 hashes together for address generation. RIPEMD-160 is used to further shorten the hash value output of SHA-256 to 160 bits.

SHA-3

In 2015, the Keccak (pronounced as "ket-chak") algorithm was standardized by the NIST as the SHA-3. Note that the purpose was not really to replace the SHA-2 standard, but to complement and coexist with it, though one can choose SHA-3 over SHA-2 in some situations.

Since both SHA-1 and SHA-2 were based on Merkle-Damgård construction, a different approach to hash function was desirable. So, not using Merkle-Damgård construction was one of the criteria set by the NIST. This was because the new design should not suffer from the limitations of Merkle-Damgård construction such as multicollision. Keccak, which became SHA-3, used a different construction method called sponge construction.

In order to make it backward compatible, it was required that SHA-3 should be able to produce variable length outputs such as 224, 256, 384, and 512 bits and also other arbitrary length outputs. This way SHA-3 became a family of cryptographic hash functions such as SHA3-224, SHA3-256, SHA3-384, SHA3 -512, and two extendable-output functions (XOFs), called SHAKE128 and SHAKE256. Also, SHA-3 had to have a tunable parameter (capacity) to allow a tradeoff between security and performance. Since SHAKE128 and SHAKE256 are XOFs, their output can be extended to any desired length, hence the name.

The following diagram (Figure 2-10) shows how SHA-3 (Keccak algorithm) is designed at a high level.

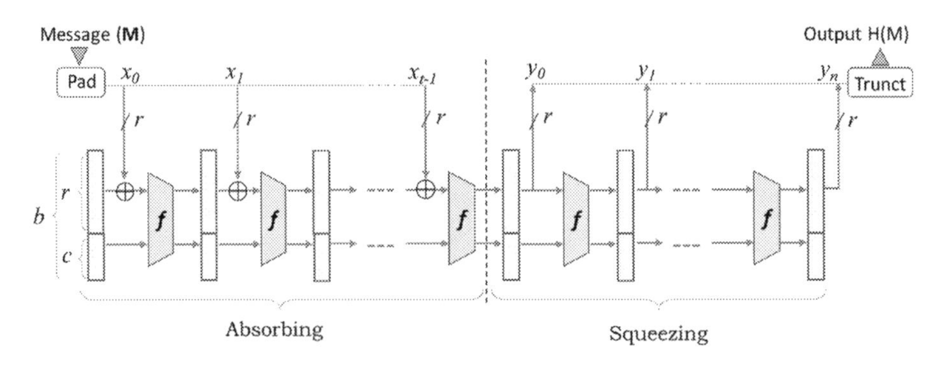

Figure 2-10. *Sponge construction for SHA-3*

A series of steps that take place for SHA-3 are as follows:

- As you can see in Figure 2-10, the message is first divided into blocks ($\mathbf{x_i}$) of size ***r*** bits. If the input data is not a multiple of ***r*** bits, then padding is required. If you are wondering about this **r**, do not worry, we will get to it in a little while. Now, let us focus on how this padding happens. For a message block $\mathbf{x_i}$ which is not a multiple of **r** and has some message **m** in it, padding happens as shown in the following:

  $\mathbf{x_i} = m \,||\, P\, 1\, \{0\}^*\, 1$

  "P" is a predetermined bit string followed by 1 {0}* 1, which means a leading and trailing 1 and some number of zeros (could be no zero bits also) that can make $\mathbf{x_i}$ a multiple of **r**. Table 2-4 shows the various values of P.

Table 2-4. *Padding in SHA-3 variants*

Mode	Output Length	P	1 {0}* 1
SHA3-224	224	11001	1 {0}* 1
SHA3-256	256	11101	1 {0}* 1
SHA3-384	384	11001	1 {0}* 1
SHA3 -512	512	11101	1 {0}* 1
Variable Length (XOFs)	Arbitrary	1111	1 {0}* 1

- As you can see in Figure 2-10, there are broadly two phases to SHA-3 sponge construction: the first one is the "Absorbing" phase for input, and the second one is the "Squeezing" phase for output. In the Absorbing phase, the message blocks ($\mathbf{x_i}$) go through various operations of the algorithm and in the Squeezing phase, the output of configurable length is computed. Notice that for both of these phases, the same function called "Kecaak-f" is used.

- For the computation of SHA3-224, SHA3-256, SHA3-384, SHA3 -512, which is effectively a replacement of SHA-2, only the first bits of the first output block y_0 are used with required level of truncation.
- The SHA-3 is designed to be tunable for its security strength, input, and output sizes with the help of tuning parameters.
- As you can see in the diagram, "b" represents the width of the state and requires that $r + c = b$. Also, "b" depends on the exponent "ℓ" such that $b = 25 \times 2^{\ell}$
- Since "ℓ" can take on values between 0 and 6, "b"can have widths {25, 50, 100, 200, 400, 800 and 1600}. It is advisable not to use the smallest two values of "b" in practice as they are just there to analyze and perform cryptanalysis on the algorithm.
- In the equation $r + c = b$, the "r" that we see is what we used to preprocess the message and divided into blocks of length "r." This is called the "bit rate." Also, the parameter "c" is called the capacity that just has to satisfy the condition $r + c = b \in \{25, 50, 100, 200, 400, 800, 1600\}$ and get computed. This way "r" and "c" are used as tuning parameters to trade off between security and performance.
- For SHA-3, the exponent value ℓ is fixed to be "6," so the value of b is 1600 bits. For this given $b = 1600$, two bit-rate values are permissible: $r = 1344$ and $r = 1088$. This leads to two distinct values of "c." So, for $r = 1344$, $c = 256$ and for $r = 1088$, $c = 512$.

- Let us now look at the core engine of this algorithm, i.e. Keccak-*f*, which is also called "Keccak-*f* Permutation." There are "n" rounds in each Keccak-*f*, where "n" is computed as: $n = 12 + 2\ell$. Since the value of ℓ is 6 for SHA-3, there will be 24 rounds in each Keccak-f. Every round takes "b" bits (r + c) input and produces the same number of "b" bits as output.
- In each round, the input "b" is called a state. This state array "b" can be represented as a three-dimensional (3-D) array $b = (5 \times 5 \times w)$, where word size $w = 2^{\ell}$. So, $w = 64$ bits, which means $5 \times 5 = 25$ words of 64 bits each. Recall that $\ell = 6$ for SHA-3, so $b = 5 \times 5 \times 64 = 1600$. The 3-D array can be shown as in Figure 2-11.

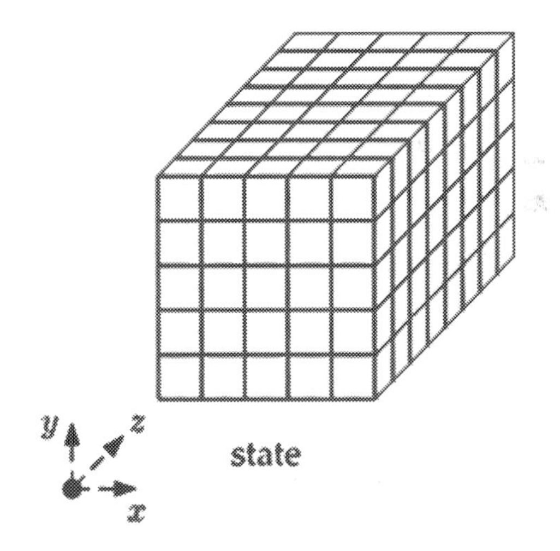

Figure 2-11. *State array representationin SHA-3*

- Each round consists of a sequence of five steps and the state array gets manipulated in each of those steps as shown in Figure 2-12.

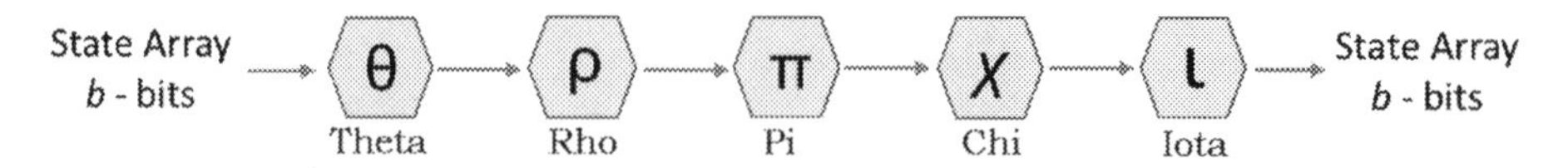

***Figure 2-12.** The five steps in each SHA-3 round*

- Without getting into much detail into each of the five steps, let us quickly learn what they do at a high level:
    - Theta (θ) step: It performs the XOR operation to provide minor diffusion.
    - Rho (ρ) step: It performs bitwise rotation of each of the 25 words.
    - Pi (π) step: It performs permutation of each of the 25 words.
    - Chi (χ) step: In this step, bits are replaced by combining those with their two subsequent bits in their rows.
    - Iota (ι) step: It XORs a round constant into one word of the state to break the symmetry.
- The last round of Keccak-*f* produces the y_0 output, which is enough for SHA-2 replacement mode, i.e., the output with 224, 256, 384, and 512 bits. Note that the least significant bits of y_0 are used for the desired length output. In case of variable length output, along with y_0, other output bits of y_1, y_2, y_3... can also be used.

When it comes to the real-life implementation of SHA-3, it is found that its performance is good in software (though not as good as SHA-2) and is excellent in hardware (better than SHA-2).

Applications of Hash Functions

The cryptographic hash functions have many different usages in different situations. Following are a few example use cases:

- Hash functions are used in verifying the integrity and authenticity of information.
- Hash functions can also be used to index data in hash tables. This can speed up the process of searching. Instead of the whole data, if we search based on the hashes (assuming the much shorter hash value compared with the whole data), then it should obviously be faster.
- They can be used to securely authenticate the users without storing the passwords locally. Imagine a situation where you do not want to store passwords on the server, obviously because if an adversary hacks on to the server, they cannot get the password from their stored hashes. Every time a user tries to log in, hash of the punched in password is calculated and matched against the stored hash. Secured, isn't it?
- Since hash functions are one-way functions, they can be used to implement PRNG.
- Bitcoin uses hash functions as a proof of work (PoW) algorithm. We will get into the details of it when we hit the Bitcoin chapter.

- Bitcoin also uses hash functions to generate addresses to improve security and privacy.
- The two most important applications are digital signatures and in MACs such as hash-based message authentication codes (HMACs).

Understanding the working and the properties of the hash functions, there can be various other use cases where hash functions can be used.

Note The Internet Engineering Task Force (IETF) adopted version 3.0 of the SSL (SSLv3) protocol in 1999, renamed it to Transport Layer Security (TLS) version 1.0 (TLSv1) protocol and defined it in RFC 2246. SSLv3 and TLSv1 are compatible as far as the basic operations are concerned.

Code Examples of Hash Functions

Following are some code examples of different hash functions. This section is just intended to give you a heads-up on how to use the hash functions programatically. Code examples are in Python but would be quite similar in different languages; you just have to find the right library functions to use.

```
# -*- coding: utf-8 -*-
import hashlib
# hashlib module is a popular module to do hashing in python

#Constructors of md5(), sha1(), sha224(), sha256(), sha384(),
and sha512() present in hashlib
md=hashlib.md5()
md.update("The quick brown fox jumps over the lazy dog")
print md.digest()
```

```
print "Digest Size:", md.digest_size, "\n", "Block Size: ",
md.block_size

# Comparing digest of SHA224, SHA256,SHA384,SHA512
print "Digest SHA224", hashlib.sha224("The quick brown fox
jumps over the lazy dog").hexdigest()
print "Digest SHA256", hashlib.sha256("The quick brown fox
jumps over the lazy dog").hexdigest()
print "Digest SHA384", hashlib.sha384("The quick brown fox
jumps over the lazy dog").hexdigest()
print "Digest SHA512", hashlib.sha512("The quick brown fox
jumps over the lazy dog").hexdigest()
# All hashoutputs are unique

# RIPEMD160 160 bit hashing example
h = hashlib.new('ripemd160')
h.update("The quick brown fox jumps over the lazy dog")
h.hexdigest()

#Key derivation Alogithm:
#Native hashing algorithms are not resistant against brutefore
attack.
#Key deviation algorithms are used for securing password
hashing.
import hashlib, binascii
algorithm='sha256'
password='HomeWifi'
salt='salt' # salt is random data that can be used as an
additional input to a one-way function
nu_rounds=1000
key_length=64 #dklen is the length of the derived key
dk = hashlib.pbkdf2_hmac(algorithm,password, salt, nu_rounds,
dklen=key_length)
```

```
print 'derieved key: ',dk
print 'derieved key in hexadeximal :', binascii.hexlify(dk)

# Check properties for hash
import hashlib

input = "Sample Input Text"
for i in xrange(20):
    # add the iterator to the end of the text
    input_text = input + str(i)
    # show the input and hash result
    print input_text, ':',  hashlib.sha256(input_text).
    hexdigest()
```

MAC and HMAC

HMAC is a type of MAC (message authentication code). As the name suggests, a MAC's purpose is to provide message authentication using Symmetric Key and message integrity using hash functions. So, the sender sends the MAC along with the message for the receiver to verify and trust it. The receiver already has the key *K* (as symmetric key cryptography is being used, so both sender and receiver have agreed on it already); they just use it to compute the MAC of the message and check it against the MAC that was sent along with the message.

In its simplest form, MAC = H(key || message). HMAC is actually a technique to turn the hash functions into MACs. In HMAC, the hash functions can be applied multiple times along with the key and its derived keys. HMACs are widely used in RFID-based systems, TLS, etc. In SSL/TLS (HTTPS is TTP within SSL/TLS), HMAC is used to allow client and server to verify and ensure that the exchanged data has not been altered during transmission. Let us take a look at a few of the important MAC strategies that are widely used:

- **MAC-then-Encrypt**: This technique requires the computation of MAC on the cleartext, appending it to the data, and then encrypting all of that together. This scheme does not provide integrity of the ciphertext. At the receiving end, the message decryption has to happen first to be able to check the integrity of the message. It ensures the integrity of the plaintext, however. TLS uses this scheme of MAC to ensure that the client-server communication session is secured.
- **Encrypt-and-MAC**: This technique requires the encryption and MAC computation of the message or the cleartext, and then appending the MAC at the end of the encrypted message or ciphertext. Notice that MAC is computed on the cleartext, so integrity of the cleartext can be assured but not of the ciphertext, which leaves scope for some attacks. Unlike the previous scheme, integrity of the cleartext can be verified. SSH (Secure Shell)uses this MAC scheme.
- **Encrypt-then-MAC**: This technique requires that the cleartext needs to be encrypted first, and then compute the MAC on the ciphertext. This MAC of the ciphertext is then appended to the ciphertext itself. This scheme ensures integrity of the ciphertext, so it is possible to first check the integrity and if valid then decrypt it. It easily filters out the invalid ciphertexts, which makes it efficient in many cases. Also, since MAC is in ciphertext, in no way does it reveal information about the plaintext. It is usually the most ideal of all schemes and has wider implementations. It is used in IPsec.

Asymmetric Key Cryptography

Asymmetric key cryptography, also known as "public key cryptography," is a revolutionary concept introduced by Diffie and Hellman. With this technique, they solved the problem of key distribution in a symmetric cryptography system by introducing digital signatures. Note that asymmetric key cryptography does not eliminate the need for symmetric key cryptography. They usually complement each other; the advantages of one can compensate for the disadvantages of the other.

Let us see a practical scenario to understand how such a system would work. Assume that Alice wants to send a message to Bob confidentially so that no one other than Bob can make sense of the message, then it would require the following steps:

Alice—The Sender:

- Encrypt the plaintext message **m** using encryption algorithm **E** and the public key $\mathbf{Puk_{Bob}}$ to prepare the ciphertext **c**.
- $\mathbf{c = E(Puk_{Bob}, m)}$
- Send the ciphertext c to Bob.

Bob—The Receiver:

- Decrypt the ciphertext **c** using decryption algorithm **D** and its private key $\mathbf{Prk_{Bob}}$ to get the original plaintext **m**.
- $\mathbf{m = D(Prk_{Bob}, c)}$

Such a system can be represented as shown in Figure 2-13.

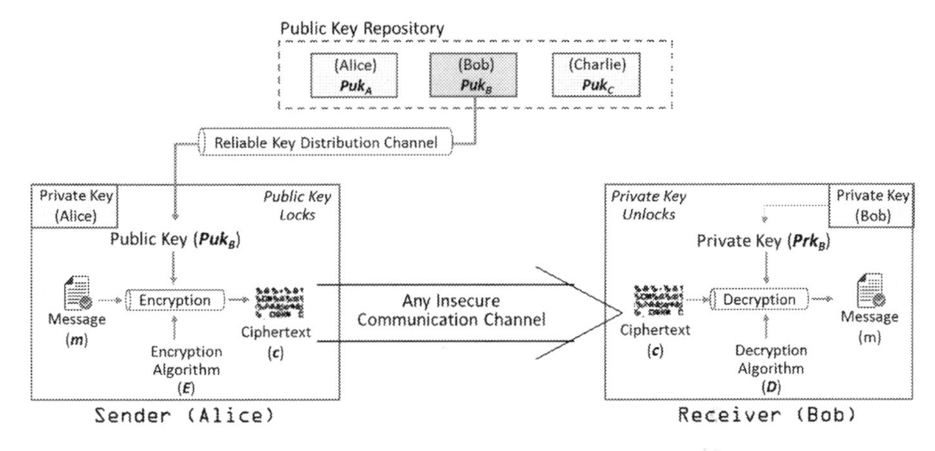

***Figure 2-13.** Asymmetric cryptography for confidentiality*

Notice that the public key should be kept in a public repository accessible to everyone and the private key should be kept as a well-guarded secret. Public key cryptography also provides a way of authentication. The receiver, Bob, can verify the authenticity of the origin of the message m in the same way.

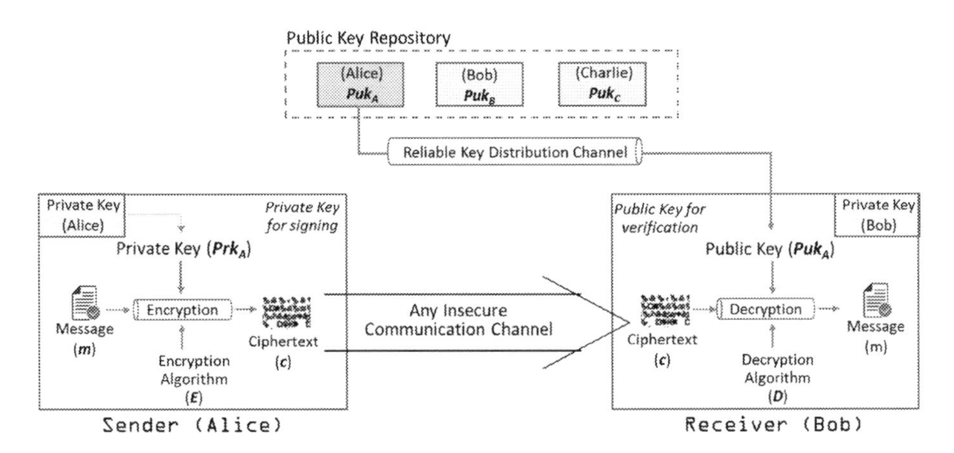

***Figure 2-14.** Asymmetric cryptography for authentication*

In the example in Figure 2-14, the message was prepared using Alice's private key, so it could be ensured that it only came from Alice. So, the entire message served as a digital signature. Note that both confidentiality and authentication are desirable. To facilitate this, public key encryption has to be used twice. The message should first be encrypted with the sender's private key to provide a digital signature. Then it should be encrypted with the receiver's public key to provide confidentiality. It can be represented as:

- $\mathbf{c = E[Puk_{Bob}, E(Prk_{Alice}, m)]}$
- $\mathbf{m = D[Puk_{Alice}, D(Prk_{Bob}, c)]}$

As you can see, the decryption happens in just its reverse order. Notice that the public key cryptography is used four times here: twice for encryption and twice for decryption. It is also possible that the sender may sign the message by applying the private key to just a small block of data derived from the message to be sent, and not to the whole message. In the real world, App stores such as Google Play or Apple App Store require that the software apps should be digitally signed before they get published.

We looked at the uses of the two keys in asymmetric cryptography, which can be summarized as follows:

- Public keys are known and accessible to everyone. They can be used to encrypt the message or to verify the signatures.
- Private keys are extremely private to individuals. They are used to decrypt the message or to create signatures.

In asymmetric or public key cryptography, there is no key distribution problem, as exchanging the agreed upon key is no longer needed. However, there is a significant challenge with this approach. How would one ensure that the public key they are using to encrypt the message is really the public key of the intended recipient and not of an intruder or eavesdropper? To solve this, the notion of a trusted third party called public key infrastructure (PKI) is introduced. Through PKIs, the authenticity of public keys is assured by the process of attestation or notarization of user identity. The way PKIs operate is that they provide verified public keys by embedding them in a security certificate by digitally signing them.

The public key encryption scheme can also be called one-way function or a trapdoor function. This is because encrypting a plaintext using the public key "Puk" is easy, but the other direction is practically impossible. No one really can deduce the original plaintext from the encrypted ciphertext without knowing the secret or private key "Prk," which is actually the trapdoor information. Also, in the context of just the keys, they are mathematically related but it is computationally not feasible to find one from the other.

We discussed the important objectives of public key cryptography such as key establishment, authentication and non-repudiation through digital signatures, and confidentiality through encryption. However, not all public key cryptography algorithms may provide all these three characteristics. Also, the algorithms are different in terms of their underlying computational problem and are classified accordingly. Certain algorithms such as RSA are based on integer factorization scheme because it is difficult to factor large numbers. Certain algorithms are based on the discrete logarithm problems in finite fields such as Diffie-Hellman key exchange (DH) and DSA. A generalized version of discrete logarithm problems is elliptic curve (EC) public key schemes. The Elliptic Curve Digital Signature Algorithm (ECDSA) is an example of it. We will cover most of these algorithms in the following section.

RSA

RSA algorithm, named after Ron Rivest, Adi Shamir, and Leonard Adleman is possibly one of the most widely used cryptographic algorithms. It is based on the practical difficulty of factoring very large numbers. In RSA, plaintext and ciphertext are integers between **0** and **n – 1** for some **n**.

We will discuss the RSA scheme from two aspects. First is generation of key pairs and second, how the encryption and decryption works. Since modular arithmetic provides the mechanism for key generation, let us quickly look at it.

Modular Arithmetic

Let **m** be a positive integer called modulus. Two integers **a** and **b** are congruent modulo **m** if:

$\mathbf{a} \equiv \mathbf{b} \pmod{\mathbf{m}}$, which implies $\mathbf{a} - \mathbf{b} = \mathbf{m} \cdot \mathbf{k}$ for some integer **k**.

Example: if $\mathbf{a} \equiv 16 \pmod{10}$ then **a** can have the following solutions:

$\mathbf{a} = \ldots, -24, -14, -4, 6, 16, 26, 36, 46$

Any of these numbers subtracted by 16 is divisible by 10. For example, $-24 - 16 = -40$, which is divisible by 10. Note that $\mathbf{a} \equiv 36 \pmod{10}$ can also have the same solutions of **a**.

As per the Quotient-Remainder theorem, only a unique solution of "**a**" exists that satisfies the condition: $\mathbf{0} \leq \mathbf{a} < \mathbf{m}$. In the example $\mathbf{a} \equiv 16 \pmod{10}$, only the value 6 satisfies the condition $0 \leq 6 < 10$. This is what will be used in the encryption/decryption process of RSA algorithm.

Let us now look at the Inverse Midulus. If **b** is an inverse to **a** modulo **m**, then it can be represented as:

$\mathbf{a\,b} \equiv \mathbf{1} \pmod{\mathbf{m}}$, which implies that $\mathbf{a\,b} - \mathbf{1} = \mathbf{m} \cdot \mathbf{k}$ for some integer **k**.

Example: 3 has inverse 7 modulo 10 since

$3 \cdot 7 = 1 \pmod{10} \Rightarrow 21 - 1 = 20$, which is divisible by 10.

Generation of Key Pairs

As discussed already, a key pair of private and public keys is needed for any party to participate in asymmetric crypto-communication. In the RSA scheme, the public key consists of **(e, n)** where **n** is called the modulus and **e** is called the public exponent. Similarly, the private key consists of **(d, n)**, where **n** is the same modulus and **d** is the private exponent.

Let us see how these keys get generated along with an example:

- Generate a pair of two large prime numbers **p** and **q**. Let us take two small prime numbers as an example here for the sake of easy understanding. So, let the two primes be **p** = 7 and **q** = 17.
- Compute the RSA modulus (**n**) as **n** = **pq**. This **n** should be a large number, typically a minimum of 512 bits. In our example, the modulus (**n**) = **pq** = 119.
- Find a public exponent **e** such that **1** < **e** < (**p** – **1**) (**q** – **1**) and there must be no common factor for **e** and (**p** – **1**) (**q** – **1**) except **1**. It implies that **e** and (**p** – **1**) (**q** – **1**) are coprime. Note that there can be multiple values that satisfy this condition and can be taken as **e**, but any one should be taken.
- In our example, (**p** – 1) (**q** – 1) = 6 × 16 = 96. So, **e** can be relatively prime to and less than 96. Let us take **e** to be 5.
- Now the pair of numbers (**e**, **n**) form the public key and should be made public. So, in our example, the public key is (5, 119).
- Calculate the private exponent **d** using **p**, **q**, and **e** considering the number **d** is the inverse of **e** modulo (**p** – 1) (**q** – 1). This implies that **d** when multiplied by **e** is equal to **1** modulo (**p** – **1**) (**q** – **1**) and **d** < (**p** – **1**) (**q** – **1**). It can be represented as:

  e d = **1** mod (**p** – **1**) (**q** – **1**)
- Note that this multiplicative inverse is the link between the private key and the public key. Though the keys are not derived from each other, there is a relation between them.

- In our example, we have to find **d** such that the above equation is satisfied. Which means, **5 d** = **1** mod **96** and also **d** < **96**.
- Solving for multiple values of **d** (can be calculated using the extended version of Euclid's algorithm), we can see that **d** = 77 satisfies our condition. See the math: 77 × 5 = 385 and 385 – 1 = 384 is divisible by 96 because 4 × 96 + 1 = 385
- We can conclude that the in our example, the private key will be (77, 119).
- Now you have got your key pairs!

Encryption/Decryption Using Key Pair

Once the keys are generated, the process of encryption and decryption are fairly simple. The math behind them is as follows:

Encrypting the plaintext message **m** to get the ciphertext message **c** is as follows:

c = **m . e** (mod **n**) given the public key (**e**, **n**) and the plaintext message **m**.

Decrypting the ciphertext message **c** to get the plaintext message **m** is as follows:

m = **c . d** (mod **n**) given the private key (**d**, **n**) and the ciphertext **c**.

Note that RSA scheme is a block cipher where the input is divided into small blocks that the RSA algorithm can consume. Also, the plaintext and the ciphertext are all integers from **0** to **n** – **1** for some integer **n** that is known to both sender and receiver. This means that the input plaintext is represented as integer, and when that goes through RSA and becomes ciphertext, they are again integers but not the same ones as input; we encrypted them remember? Now, considering the same key pairs from the previous example, let us go through the steps to understand how it works practically:

- The sender wants to send a text message to the receiver whose public key is known and is say (**e**, **n**).
- The sender breaks the text message into blocks that can be represented as a series of numbers less than **n**.
- The ciphertext equivalents of plaintext can be found using **c** = **m e** (mod **n**). If the plaintext (**m**) is 19 and the public key is (5, 119) with **e** = 5 and **n** = 119, then the ciphertext **c** will be 195(mod 119) = 2, 476, 099 (mod 119) = 66, which is the remainder and 20,807 is the quotient, which we do not use. So, **c** = 66
- When the ciphertext 66 is received at the receiver's end, it needs to be decrypted to get the plaintext using **m** = **c d** (mod **n**).
- The receiver already has the private key (**d**, **n**) with **d** = 77 and **n** = 119, and received the ciphertext **c** = 66 by the sender. So, the receiver can easily retrieve the plaintext using these values as **m** = 6,677(mod 119) = 19
- For the modular arithmetic calculations, there are many online calculators that you can play around with, such as: `http://comnuan.com/cmnn02/cmnn02008/`

We looked at the math behind RSA algorithm. Now we know that **n** (supposed to be a very large number) is publicly available. Though it is public, factoring this large number to get the prime numbers **p** and **q** is extremely difficult. The RSA scheme is based on this practical difficulty of factoring large numbers. If **p** and **q** are not large enough, or the public key **e** is small, then the strength of RSA goes down. Currently, RSA keys are typically between 1024 and 2048 bits long. Note that the computational overhead of the RSA cryptography increases with the size of the keys.

In situations where the amount of data is huge, it is advisable to use a symmetric encryption technique and share the key using an asymmetric encryption technique such as RSA. Also, we looked at one of the aspects of RSA, that is, for encryption and decryption. However, it can also be used for authentication through digital signature. Just to give a high-level idea, one can take the hash of the data, sign it using their own private key, and share it along with the data. The receiver can check with the sender's public key and ensure that it was the sender who sent the data, and not someone else. This way, in addition to secure key transport, the public key encryption method RSA also offers authentication using a digital signature. Note here that a different algorithm called digital signature algorithm (DSA) can also be used in such situations that we will learn about in the following section.

RSA is widely being used with HTTPS on web browsers, emails, VPNs, and satellite TV. Also, many commercial applications or the apps in app stores are also digitally signed using RSA. SSH also uses public key cryptography; when you connect to an SSH server, it broadcasts a public key that can be used to encrypt data to be sent to that server. The server can then decrypt the data using its private key.

Digital Signature Algorithm

The DSA was designed by the NSA as part of the Digital Signature Standard (DSS) and standardized by the NIST. Note that its primary objective is to sign messages digitally, and not encryption. Just to paraphrase, RSA is for both key management and authentication whereas DSA is only for authentication. Also, unlike RSA, which is based on large-number factorization, DSA is based on discrete logarithms. At a high level, DSA is used as shown in Figure 2-15.

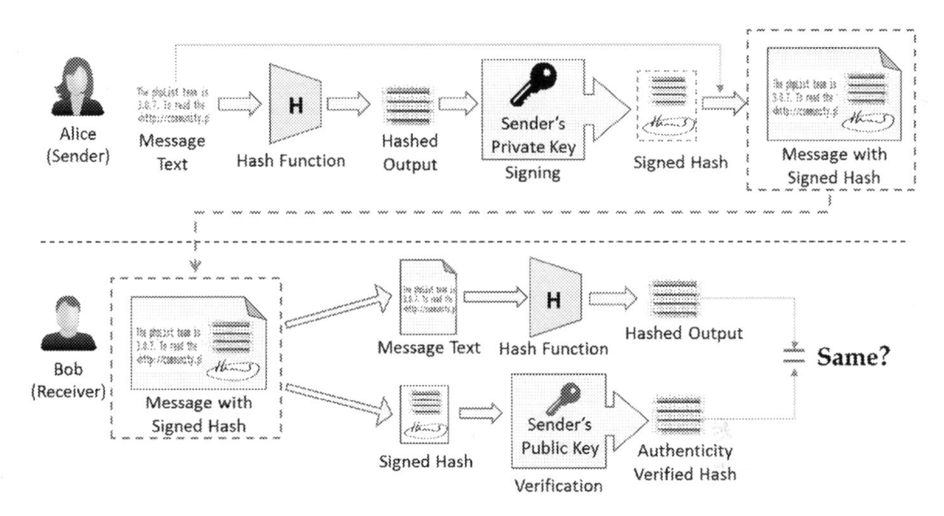

***Figure 2-15.** Digital Signature Algorithm (DSA)*

As you can see in Figure 2-15, the message is first hashed and then signed because it is more secured compared with signing and then hashing it. Ideally, you would like to verify the authenticity before doing any other operation. So, after the message is signed, the signed hash is tagged with the message and sent to the receiver. The receiver can then check the authenticity and find the hash. Also, hash the message to get the hash again and check if the two hashes match. This way, DSA provides the following security properties:

- Authenticity: Signed by private key and verified by public key
- Data integrity: Hashes will not match if the data is altered.
- Non-repudiation: Since the sender signed it, they cannot deny later that they did not send the message. Non-repudiation is a property that is most desirable in situations where there are chances of a dispute over the exchange of data. For example, once an order is placed electronically, a purchaser cannot deny the purchase order if non-repudiation is enabled in such a situation.

A typical DSA scheme consists of three algorithms: (1) key generation, (3) signature generation, and (3) signature verification.

Elliptic Curve Cryptography

Elliptic curve cryptography (ECC) actually evolved from Diffie-Hellman cryptography. It was discovered as an alternative mechanism for implementing public key cryptography. It actually refers to a suite of cryptographic protocols and is based on the discrete logarithm problem, as in DSA. However, it is believed that the discrete logarithmic problem is even harder when applied to the points on an elliptic curve. So, ECC offers greater security for a given key size. A 160-bit ECC key is considered to be as secured as a 1024-bit RSA key. Since smaller key sizes in ECC can provide greater security and performance compared with other public key algorithms, it is widely used in small embedded devices, sensors, and other IoT devices, etc. There are extremely efficient hardware implementations available for ECC.

ECC is based on a mathematically related set of numbers on an elliptic curve over finite fields. Also, it has nothing to do with ellipses! Mathematically, an elliptic curve satisfies the following mathematical equation:

$y^2 = x^3 + ax + b$, where $4 a^3 + 27 b^2 \neq 0$

With different values of "**a**" and "**b**", the curve takes different shapes as shown in the following diagram:

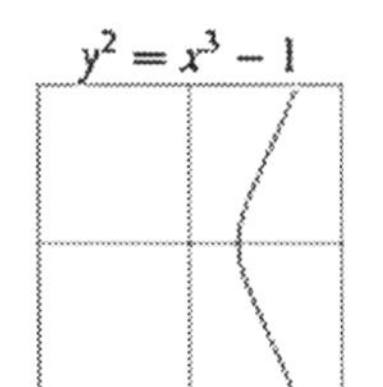

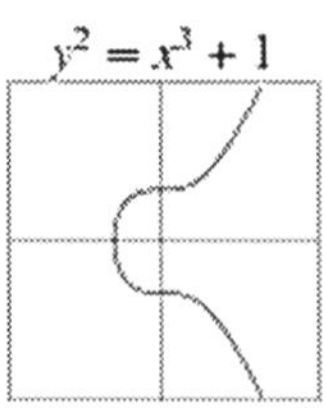

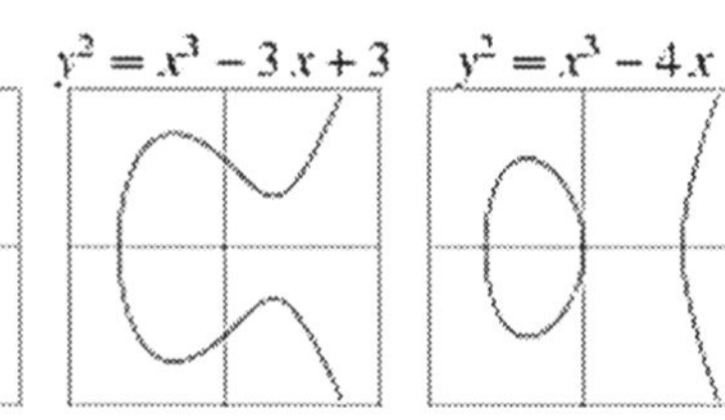

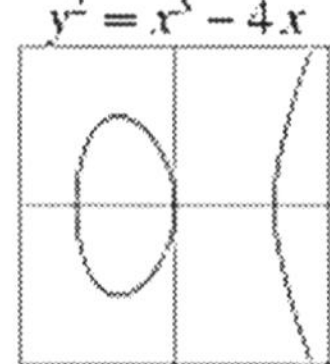

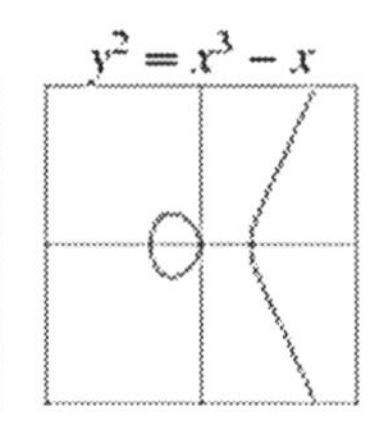

There are several important characteristics of elliptic curves that are used in cryptography, such as:

- They are horizontally symmetrical. i.e., what is below the X-axis is a mirror image of what is above the X-axis. So, any point on the curve when reflected over the X-axis still remains on the curve.
- Any nonvertical line can intersect the curve in at most three places.
- If you consider two points **P** and **Q** on the elliptic curve and draw a line through them, the line may exactly cross the curve at one more places. Let us call it (– **R**). If you draw a vertical line through (– **R**), it will cross the curve at, say, **R**, which is a reflection of the point (– **R**). Now, the third property implies that **P** + **Q** = **R**. This is called "point addition," which means adding two points on an elliptic curve will lead you to another point on the curve. Refer to the following diagram for a pictorial representation of these three properties.

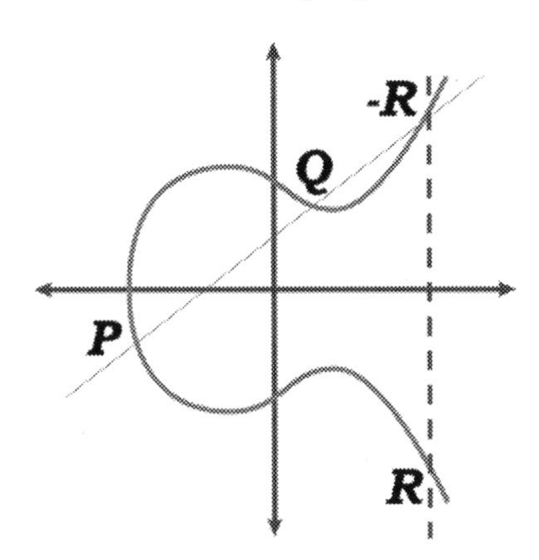

- So, you can apply point addition to any two points on the curve. Now, in the previous bullet-point, we did point addition of **P** and **Q** (**P** + **Q**) and found – **R** and then ultimately arrived at **R**. Once we arrive at **R**, we can then draw a line from **P** to **R** and see that the line intersects the graph again at a third point. We can then take that point and move along a vertical line until it intersect the graph again. This becomes the point addition for points **P** and **R**. This process with a fixed **P** and the resulting point can continue as long as we want, and we will keep getting new points on the curve.
- Now, instead of two points **P** and **Q**, what if we apply the operation to the same point **P**, i.e., **P** and **P** (called "point doubling"). Obviously, infinite numbers of lines are possible through P, so we will only consider the tangential line. The tangent line will cross the curve in one more point and a vertical line from there will cross the curve again to get to the final value. It can be shown as follows:

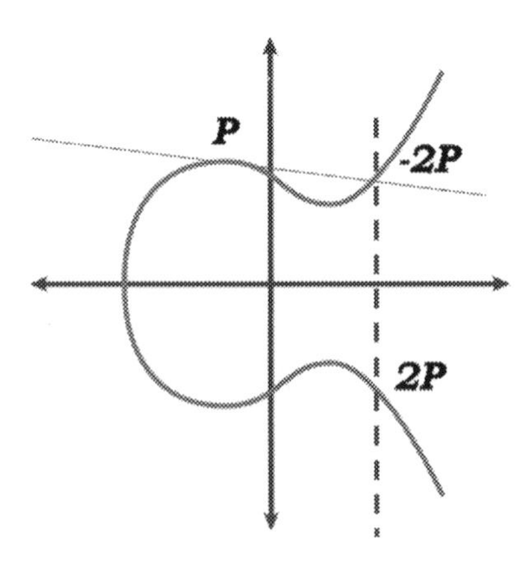

- It is evident that we can apply point doubling "**n**" number of times to the initial point and every time it will lead us to a different point on the curve. The first time we applied point doubling to the point **P**, it took us to the resulting point **2P** as you can see in the diagram. Now, if the same is repeated "**n**" number of times, we will reach a point on the curve as shown in the following diagram:

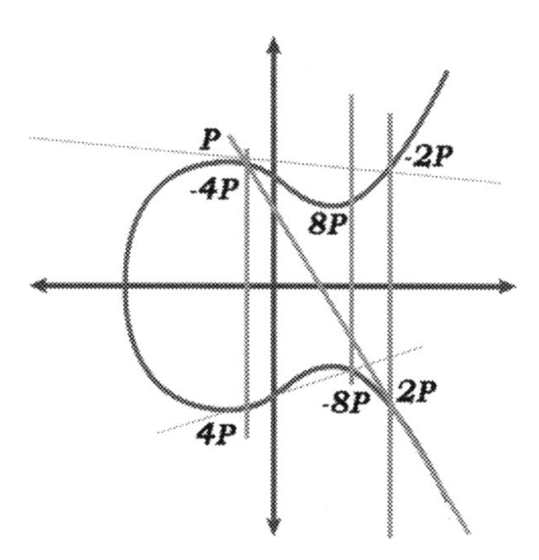

- In the aforementioned scenario, when the initial and final point is given, there is no way one can say that the point doubling was applied "**n**" number of times to reach the final resulting point except trying for all possible "**n**" one by one. This is the discrete logarithm problem for ECC, where it states that given a point **G** and **Q**, where **Q** is a multiple of **G**, find "**d**" such that **Q** = **d G**. This forms the one-way function with no shortcuts. Here, **Q** is the public key and **d** is the private key. Can you extract private key **d** from public key **Q**? This is the elliptic curve discrete logarithm problem, which is computationally difficult to solve.

- Further to this, the curve should be defined over a finite field and not take us to infinity! This means the "max" value on the X-axis has to be limited to some value, so just roll the values over when we hit the maximum. This value is represented as **P** (not the **P** used in the graphs here) in the ECC cryptosystem and is called "modulo" value, and it also defines the key size, hence the finite field. In many implementations of ECC, a prime number for "**P**" is chosen.
- Increased size of "**P**" results in more usable values on the curve, hence more security.
- We observed that point addition and point doubling form the basis for finding the values that are used for encryption and decryption.

So, in order to define an ECC, the following domain parameters need to be defined:

- The Curve Equation: $y^2 = x^3 + ax + b$, where $4a^3 + 27b^2 \neq 0$
- **P**: The prime number, which specifies the finite field that the curve will be defined over (modulo value)
- **a** and **b**: Coefficients that define the elliptic curve
- **G**: Base point or the generator point on the curve. This is the point where all the point operations begin and it defines the cyclic subgroup.
- **n**: The number of point operations on the curve until the resultant line is vertical. So, it is the order of **G**, i.e., the smallest positive number such that $nG = \infty$. It is normally prime.
- **h**: It is called "cofactor," which is equal to the order of the curve divided by **n**. It is an integer value and usually close to 1.

Note that ECC is a great technique to generate the keys, but is used alongside other techniques for digital signatures and key exchange. For example, Elliptic Curve Diffie-Hellman (ECDH) is quite popularly used for key exchange and ECDSA is used for digital signatures.

Elliptic Curve Digital Signature Algorithm

The ECDSA is a type of DSA that uses ECC for key generation. As the name suggests, its purpose is digital signature, and not encryption. ECDSA can be a better alternative to RSA in terms of smaller key size, better security, and higher performance. It is one of the most important cryptographic components used in Bitcoins!

We already looked at how digital signatures are used to establish trust between the sender and receiver. Since authenticity of the sender and integrity of the message can be verified through digital signatures, two unknown parties can transact with each other. Note that the sender and the receiver have to agree on the domain parameters before engaging in the communication.

There are broadly three steps to ECDSA: key generation, signature generation, and signature verification.

Key Generation

Since the domain parameters (**P**, **a**, **b**, **G**, **n**, **h**) are preestablished, the curve and the base point are known by both parties. Also, the prime **P** that makes it a finite field is also known (**P** is usually 160 bits and can be greater as well). So, the sender, say, Alice does the following to generate the keys:

- Select a random integer **d** in the interval [**1**, **n** – **1**]
- Compute **Q** = **d G**
- Declare **Q** is the public key and keep **d** as the private key.

Signature Generation

Once the keys are generated, Alice, the sender, would use the private key "**d**" to sign the message (**m**). So, she would perform the following steps in the order specified to generate the signature:

- Select a random number **k** in the interval [**1**, **n** – **1**]
- Compute **k.G** and find the new coordinates (x_1, y_1) and find $\mathbf{r} = x_1 \bmod \mathbf{n}$

  If **r** = 0, then start all over again
- Compute **e** = SHA-1 (**m**)
- Compute $\mathbf{s} = \mathbf{k}^{-1} (\mathbf{e} + \mathbf{d} \,.\, \mathbf{r}) \bmod \mathbf{n}$

  If s = 0, then start all over again from the first step
- Alice's signature for the message (**m**) would now be (**r**, **s**)

Signature Verification

Let us say Bob is the receiver here and has access to the domain parameters and the public key **Q** of the sender Alice. As a security measure, Bob should first verify that the data he has, which is the domain parameters, the signature, and Alice's public key **Q** are all valid. To verify Alice's signature on the message (**m**), Bob would perform the following operations in the order specified:

- Verify that **r** and **s** are integers in the interval [**1**, **n** – **1**]
- Compute **e** = SHA-1 (**m**)
- Compute $\mathbf{w} = \mathbf{s}^{-1} \bmod \mathbf{n}$
- Compute $\mathbf{u}_1 = \mathbf{e}\ \mathbf{w} \bmod \mathbf{n}$, and $\mathbf{u}_2 = \mathbf{r}\ \mathbf{w} \bmod \mathbf{n}$
- Compute $\mathbf{X} = \mathbf{u}_1\, G + \mathbf{u}_2\, G$, where X represents the coordinates, say (x_2, y_2)

- Compute **v** = x_1 mod **n**
- Accept the signature if **r** = **v**, otherwise reject it

In this section, we looked at the math behind ECDSA. Recollect that we used a random number while generating the key and the signature. It is extremely important to ensure that the random numbers generated are actually cryptographically random. In many use cases, 160-bit ECDSA is used because it has to match with the SHA-1 hash function.

Out of so many use cases, ECDSA is used in digital certificates. In its simplest form, a digital certificate is a public key, bundled with the device ID and the certificate expiration date. This way, certificates enable us to check and confirm to whom the public key belongs and the device is a legitimate member of the network under consideration. These certificates are very important to prevent "impersonation attack" in key establishment protocols. Many TLS certificates are based on ECDSA key pair and this usage continues to grow.

Code Examples of Assymetric Key Cryptography

Following are some code examples of different public ley algorithms. This section is just intended to give you a heads-up on how to use different algorithms programatically. Code examples are in Python but would be quite similar in different languages; you just have to find the right library functions to use.

```
# -*- coding: utf-8 -*-
import Crypto
from Crypto.PublicKey import RSA
from Crypto import Random
from hashlib import sha256
```

```
# Function to generate keys with default lenght 1024
def generate_key(KEY_LENGTH=1024):
    random_value= Random.new().read
    keyPair=RSA.generate(KEY_LENGTH,random_value)
    return keyPair

#Generate Key for ALICE and BOB
bobKey=generate_key()
aliceKey=generate_key()

#Print Public Key of Alice and Bob. This key could shared
alicePK=aliceKey.publickey()
bobPK=bobKey.publickey()

print "Alice's Public Key:", alicePK
print "Bob's Public Key:", bobPK

#Alice wants to send a secret message to Bob. Lets create a
dummy message for Alice
secret_message="Alice's secret message to Bob"
print "Message",  secret_message

# Function to generate a signature
def generate_signature(key,message):
    message_hash=sha256(message).digest()
    signature=key.sign(message_hash,'')
    return signature

# Lets generate a signature for secret message
alice_sign=generate_signature(aliceKey,secret_message)

# Before sending message in network, encrypt message using the
Bob's public key...
encrypted_for_bob  = bobPK.encrypt(secret_message, 32)
```

```
# Bob decrypts secret message using his own private key...
decrypted_message   = bobKey.decrypt(encrypted_for_bob)
print "Decrypted message:", decrypted_message

# Bob will use the following function to verify the signature
from Alice using her public key
def verify_signature(message,PublicKey,signature):
    message_hash=sha256(message).digest()
    verify = PublicKey.verify(message_hash,signature)
    return verify

# bob is verifying using decrypted message and alice's public
key
print "Is alice's signature for decrypted message valid?",
verify_signature(decrypted_message,alicePK, alice_sign)
```

The ECDSA Algorithm

```
import ecdsa

# SECP256k1 is the Bitcoin elliptic curve
signingKey = ecdsa.SigningKey.generate(curve=ecdsa.SECP256k1)
# Get the verifying key
verifyingKey = signingKey.get_verifying_key()

# Generate The signature of a message
signature = signingKey.sign(b"signed message")

# Verify the signature is valid or invalid for a message
verifyingKey.verify(signature, b"signed message") # True.
Signature is valid

# Verify the signature is valid or invalid for a message
assert verifyingKey.verify(signature, b"message") # Throws an
error. Signature is invalid for message
```

Diffie-Hellman Key Exchange

We already looked at symmetric key cryptography in the previous sections. Recollect that sharing the secret between the sender and the receiver is a very big challenge. As a rule of thumb, we are now aware that the the communication channel is always insecure. There could always be an *Eve* trying to intercept your message while it is being transmitted by using various different kinds of attacks. So, the technique of DH was developed for securely exchanging the cryptographic keys. Obviously, you must be wondering how secure key exchange is possible when the communication channel itself is insecured. Well, later in this section you will see that the DH technique is not really sharing the entire secret key between two parties, rather it is about creating the key together. At the end of the day, what is important is that the sender and the receiver both have the same key. However, keep in mind that it is not asymmetric key cryptography, as encryption/decrytion does not take place during the exchange. In fact, it was the base upon which asymmetric key cryptography was later designed. The reason we are looking at this technique now is because a lot of math that we already studied in the previous section is useful here.

Let us first try to understand the concept at a high level before getting into the mathematical explanation. Take a look at the following (Figure 2-16), where a simple explanation of DH algorithm is presented with colors.

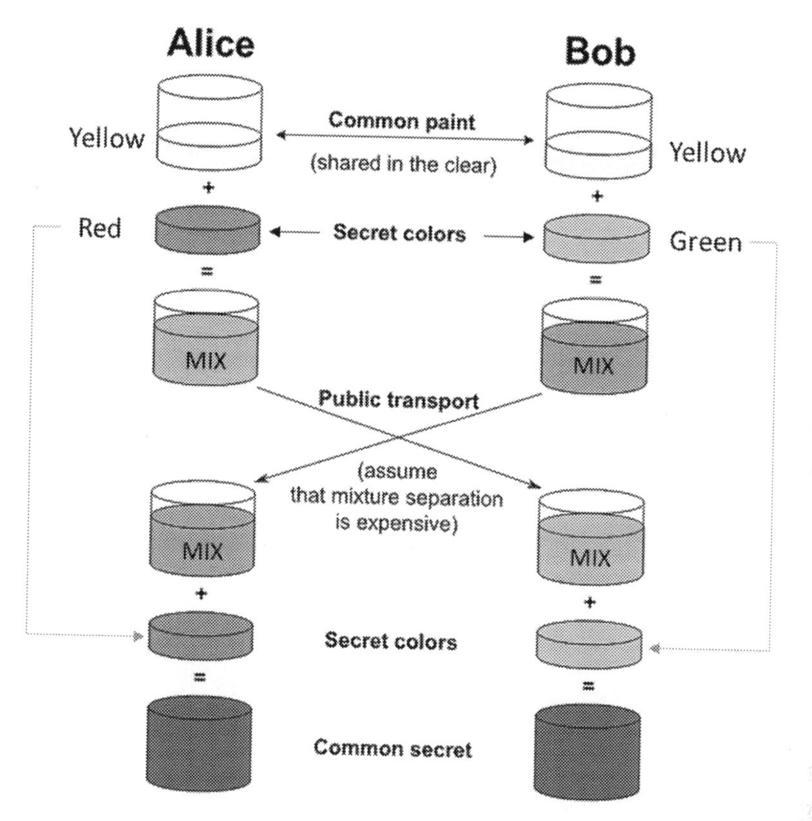

Figure 2-16. *Diffie-Hellman key exchange illustration*

Notice that only the yellow color was shared between the two parties in the first step, which may represent any other color or a randon number. Both parties then add their own secret to it and make a mixture. That mixture is again shared through the same insecured channel. Respective parties then add their secret to it and form their final common secret. In this example with colors, observe that the common secrets are the combination of same sets of colors. Let us now look at the actual mathematical steps that take place for the generation of keys:

- Alice and Bob agree on $P = 23$ and $G = 9$
- Alice chooses private key $a = 4$, computes $9^4 \bmod 23 = 6$ and sends it to Bob

- Bob chooses private key b = 3, computes 9^3 mod 23 = 16 and sends it to Alice
- Alice computes 16^4 mod 23 = 9
- Bob computes 6^3 mod 23 = 9

If you follow through these steps, you will find that both Alice and Bob are able to generate the same secret key at their ends that can be used for encryption/decryption. We used small numbers in this example for easy understanding, but large prime numbers are used in real-world use cases. To understand it better, let us go through the following code snippet and see how DH algorithm can be implemented in a simple way:

```
/* Program to calculate the Keys for two parties using Diffie-
Hellman Key exchange algorithm */

// function to return value of a ^ b mod P
long long int power(long long int a, long long int b, long long
int P)
{
    if (b == 1)
        return a;

    else
        return (((long long int)pow(a, b)) % P);
}

//Main program for DH Key computation
int main()
{
    long long int P, G, x, a, y, b, ka, kb;

    // Both the parties agree upon the public keys G and P
    P = 23; // A prime number P is taken
    printf("The value of P : %lld\n", P);
```

```
    G = 9; // A primitve root for P, G is taken
    printf("The value of G : %lld\n\n", G);

    // Alice will choose the private key a
    a = 4; // a is the chosen private key
    printf("The private key a for Alice : %lld\n", a);
    x = power(G, a, P); // gets the generated key

    // Bob will choose the private key b
    b = 3; // b is the chosen private key
    printf("The private key b for Bob : %lld\n\n", b);
    y = power(G, b, P); // gets the generated key

    // Generating the secret key after the exchange of keys
    ka = power(y, a, P); // Secret key for Alice
    kb = power(x, b, P); // Secret key for Bob


    printf("Secret key for the Alice is : %lld\n", ka);
    printf("Secret Key for the Bob is : %lld\n", kb);

    return 0;
}
```

Note While the discrete logarithm problem is traditionally used (the x^y mod p), the general process can be modified to use elliptic curve cryptography as well.

Symmetric vs. Asymmetric Key Cryptography

We looked at various aspects and types of both symmetric and asymmetric key algorithms. Obviously, their design goals and implications are different. Let us have a comparative analysis so that we use the right one at the right place.

- Symmetric key cryptography is also referred to as private key cryptography. Similarly, asymmetric key cryptography is also called public key cryptography.
- Key exchange or distribution in symmetric key cryptography is a big headache, unlike asymmetric key cryptography.
- Asymmetric encryption is quite compute-intensive because the length of the keys is usually large. Hence, the process of encryption and decryption is slower. On the contrary, symmetric encryption is faster.
- Symmetric key cryptography is appropriate for long messages because the speed of encryption/decryption is fast. Asymmetric key cryptography is appropriate for short messages, and the speed of encryption/decryption is slow.
- In symmetric key cryptography, symbols in plaintext and ciphertext are permuted or substituted. In asymmetric key cryptography, plaintext and ciphertext are treated as integers.
- In many situations, when symmetric key is used for encryption and decryption, asymmetric key technique is used to share and agree upon the key used in encryption.

- Asymmetric key cryptography finds its strongest application in untrusted environments, when parties involved have no prior relationship. Since the unknown parties do not get any prior opportunity to establish shared secret keys with each other, sharing of sensitive data is secured through public key cryptography.
- Symmetric cryptographic techniques do not provide a way for digital signatures, which are only possible through asymmetric cryptography.
- Another good case is the number of keys required among a group of nodes to communicate with each other. How many keys do you think would be needed among, say, 100 participants when symmetric key cryptography is needed? This problem of finding the keys needed can be approached as a complete graph problem with order 100. Like each vertex requires 99 connected edges to connect with everyone, every participant would need 99 keys to establish secured connections with all other nodes.

  So, in total, the keys needed would be 100 * (100 – 1)/2 = 4,950. It can be generalized for "**n**" number of participants as **n** * (**n** – 1)/2 keys in total. With an increased number of participant, it becomes a nightmare! However, in the case of asymmetric key cryptography, each participant would just need two keys (one private and one public). For a network of 100 participants, total keys needed would be just 200. Table 2-5 shows some sample data to give you an analogy on the increased number of keys needed when the number of participants increases.

Table 2-5. *Key Requirements Comparison for Symmetric and Asymmetric Key Techniques*

Number of Participants	Number of Symmetric Keys	Number of Asymmetric Keys
2	1	4
4	6	8
10	45	20
50	1225	100
100	4950	200
1000	499500	2000

Game Theory

Game Theory is a certainly quite an old concept and is being used in many real-life situations to solve complex problems. The reason we are covering this topic at a high level is because it is used in Bitcoins and many other blockchain solutions. It was formally introduced by John von Neumann to study economic decisions. Later, it was more popularized by John Forbes Nash Jr because of his theory of "Nash Equilibrium," which we will look into shortly. Let us first understand what game theory is.

Game theory is a theory on games, where the games are not just what children play. Most are situations where two or more parties are involved with some strategic behavior. Examples: A cricket tournament is a game, two conflicting parties in a court of law with lawyers and juries is a game, two siblings fighting over an ice cream is a game, a political election is a game, a traffic signal is also a game. Another example: Say you applied for a blockchain job and you are selected and offered a job offer with some salary, but you reject the offer, thinking there is a huge gap in the demand and supply and chances are good they will revise the offer with a higher salary. You must be thinking now, what is not a game? Well, in real situations, almost everything is a game. So, a "game" can be defined as a situation involving a "correlated rational choice." What it means is that the prospects available for any player are dependent not only on their own choices, but also on the choices that others make in a given situation. In other words, if your fate is impacted by the actions of others, then you are in a game. So what is game theory?

Game theory is a study of strategies involved in complex games. It is the art of making the best move, or opting for a best strategy in a given situation based on the objective. To do so, one must understand the strategy of the opponent and also what the opponent thinks your move is going to be. Let us take a simple example: There are two siblings, one elder and the other younger. Now, there are two ice creams in the fridge, one is orange flavor and the other is mango flavor. The elder one wants to eat the orange flavor, but knows if he opts for that, then the younger one would cry for the same orange. So, he opts for the mango flavored ice cream and it turns out as expected, the younger one wants the same. Now, the elder one pretends to have sacrificed the mango flavored ice cream and gives it to the younger one and eats the orange one himself. Look at the situation: this is a win-win for both the parties, as this was the objective of the elder one. If the elder one wanted, he could simply have fought with the younger kid and got the orange one if that was his objective. In the second case, the elder one would strategize where to hit so that the younger kid is not injured much but enough so that he gives up on the orange flavored ice cream. This is game theory: what is your objective and what should be your best move?

One more example: more on a business side this time. Imagine that you are a vendor supplying vegetables to a town. There are, say, three ways to get to the town, out of which one is a regular route in the sense that everyone goes by that route, maybe because it is shorter and better. One day, you see that the regular route has been blocked because of some repair activity and in no way can you go by that route. You are now left with two other routes. One of those is a short route to the destination town but is a little narrow. The other one is a little longer route but wide enough. Here, you have to make a strategy as to which route of the two you need to go by. The situation may be such that there is heavy traffic on the roads and many people would try to get through the shortest route. This can lead to heavy congestion on that route and can cause a huge delay. So, you decided to take the longer route to reach the town on time, but at the cost of few extra dollars spent on fuel. You are sure you can easily get compensated for that if you arrive on time and sell your vegetables early at a good price. This is game theory: what is your best move for the objective you have in mind, which is usually finding an optimal solution.

In many situations, the role that you play and your objective both play a vital role in formulating the strategy. Example: If you are an organizer of a sport event, and not a participant in the competition, then you would formulate a strategy where your objective could be that you want the participants to play by the rules and follow the protocol. This is because you do not care who wins at the end, you are just an organizer. On the other hand, a participant would strategize the winning moves by taking into account the strengths and weaknesses of the opponent, and the rules imposed by the organizer because there could be penalties if you break the rules. Now, let us consider this situation with you playing the role of the organizer. You should consider if there could be a situation where a participant breaks a rule and loses one point but injures the opponent so much that they cannot compete any longer. So, you have to take into account what the participants can think and set your rules accordingly.

Let us try to define game theory once again based on what we learned from the previous examples. It is the method of modeling real-life situations in the form of a game and analyzing what the best strategy or move of a person or an entity could be in a given situation for a desired outcome. Concepts from game theory are widely used in almost every aspect of life, such as politics, social media, city planning, bidding, betting, marketing, distributed storage, distributed computing, supply chains, and finance, just to name a few. Using game theoretic concepts, it is possible to design systems where the participants play by the rules without assuming emotional or moral values of them. If you want to go beyond just building a proof of concept and get your product or solution to production, then you should prioritize game theory as one of the most important elements. It can help you build robust solutions and lets you test those with different interesting scenarios. Well, many people already think in game theoretic perspectives without knowing it is game theory. However, if you are equipped with the many tools and techniques from game theory, it definitely helps.

Nash Equilibrium

In the previous section, we looked at different examples of games. There are many ways to classify games, such as cooperative/noncooperative games, symmetric/asymmetric games, zero-sum/non-zero-sum games, simultaneous/sequential games, etc. More generally, let us focus on the cooperative/noncooperative perspective here, because it is related to the Nash equilibrium.

As the name suggests, the players cooperate with each other and can work together to form an alliance in cooperative games. Also, there can be some external force applied to ensure cooperative behavior among the players. On the other hand, in noncooperative games, the players compete as individuals with no scope to form an alliance. The participants just look after their own interests. Also, no external force is available to enforce cooperative behavior.

Nash equilibrium states that, in any noncooperative games where the players know the strategies of each other, there exists at least one equilibrium where all the players play their best strategies to get the maximum profits and no side would benefit by changing their strategies. If you know the strategies of other players and you have your own strategy as well, if you cannot benefit by changing your own strategy, then this is the state of Nash equilibrium. Thus, each strategy in a Nash equilibrium is a best response to all other strategies in that equilibrium.

Note that a player may strategize to win as an individual player, but not to defeat the opponent by ensuring the worst for the opponents. Also, any game when played repeatedly may eventually fall into the Nash equilibrium.

In the following section, we will look at the "prisoner"s dilemma" to get a concrete understanding of the Nash equilibrium.

Prisoner's Dilemma

Many games in real life can also be non-zero-sum games. Prisoner's dilemma is one such example, which can be broadly categorized as a symmetric game. This is because, if you change the identities of the players (e.g., if two players "A" and "B" are playing, then "A" becomes "B" and "B" becomes "A"), and also the strategies do not change, then the payoff remains the same. This is what a symmetric game is.

Let us start directly with an example. Assume that there are two guys, Bob and Charlie, who are caught by the cops for selling drugs independently, say in different locations. They are kept in two different cells for interrogation. They were then toldd that they would be sentenced to jail for two years for this crime. Now, the cops somehow suspect that these two guys could also be involved in the robbery that just happened last week. If they did not do the robbery, then it is two years of imprisonment anyway. So, the cops have to strategize a way to get to the truth. So here is what they do.

The cops go to Bob and give him a choice, a good choice that goes like this. If Bob confesses his crime and Charlie does not, then his punishment would go down from two years to just one year and Bob gets five years. However, if Bob denies and Charlie confesses, then Bob gets five years and Charlie gets just one year. Also, if both confess, then both get three years of imprisonment. Similarly, the same choice is given to Charlie as well. What do you think they are going to do? This situation is called the prisoner's dilemma.

Both Bob and Charlie are in two different cells. They cannot talk to each other and conclude with the situation where they both deny and get two years in jail (just for the drug dealing case), which seems to be the global optimum in this situation. Well, even if they could talk to each other, they may not really trust each other.

What would go through Bob's mind now? He has two choices, confess or deny. He knows that Charlie would choose what is best for him, and he himself is no different. If he denies and Charlie confesses, then he is in trouble by getting five years of jail and Charlie gets just one year of jail. He certainly does not want to get into this situation.

If Bob confesses, then Charlie has two choices: confess or deny. Now Bob thinks that if he confesses, then whatever Charlie does, he is not getting more than three years. Let us state these scenarios for Bob.

- Bob confesses and Charlie denies—Bob gets one year, Charlie gets five years (best case given Bob confesses)
- Bob confesses and Charlie also confesses—Both Bob and Charlie get three years (worst case given Bob confesses)

This situation is called Nash equilibrium where each party has taken the best move, given the choices of the other party. This is definitely not the global optimum, but represents the best move as an individual. Now, if you look at this situation as an outsider, you would say both should deny and get two years. But when you play as a participant in the game, Nash equilibrium is what you would eventually fall into. Note that this is the most stable stage where you changing your decision does not benefit you at all. It can be pictorially represented as shown in Figure 2-17.

		Charlie	
		Confess	Deny
Bob	Confess	3, 3	1, 5
	Deny	5, 1	2, 2

Figure 2-17. *Prisoner's dilemma–payoff matrix*

Byzantine Generals' Problem

In the previous section, we looked at different examples of games and learned a few game theory concepts. Now we will discuss a specific problem from the olden days that is still widely used to solve many computer science as well as real-life problems.

The Byzantine Generals' Problem was a problem faced by the Byzantine army while attacking a city. The situation was straightforward yet very difficult to deal with. To put it simply, the situation was that several army factions commanded by separate generals surrounded a city to win over it. The only chance of victory is when all the generals attack the city together. However, the problem is how to reach a consensus. This implies that either all the generals should attack or all of them should retreat. If some of them attack and some retreat, then chances are greater they would lose the battle. Let us take an example with numbers to be able to understand the situation better.

Let us assume a situation where there are five factions of the Byzantine army surrounding a city. They would attack the city if at least three out of five generals are willing to attack, but retreat otherwise. If there is a traitor among the generals, what he can do is vote for attack with the generals willing to attack and vote for retreat with the generals willing to retreat. He can do so because the army is dispersed in factions, which makes centralized coordination difficult. This can result in two generals attacking the city and getting outnumbered and defeated. There could be more complicated issues with such a situation:

- What if there is more than one traitor?
- How would the message coordination between generals take place?
- What if a messenger is caught/killed/bribed by the city commander?
- What if a traitor general forges a different message and fools other generals?
- How to find the generals who are honest and who are traitors?

As you can see, there are so many challenges that need to be addressed for a coordinated attack on the city. It can be pictorially represented as in Figure 2-18.

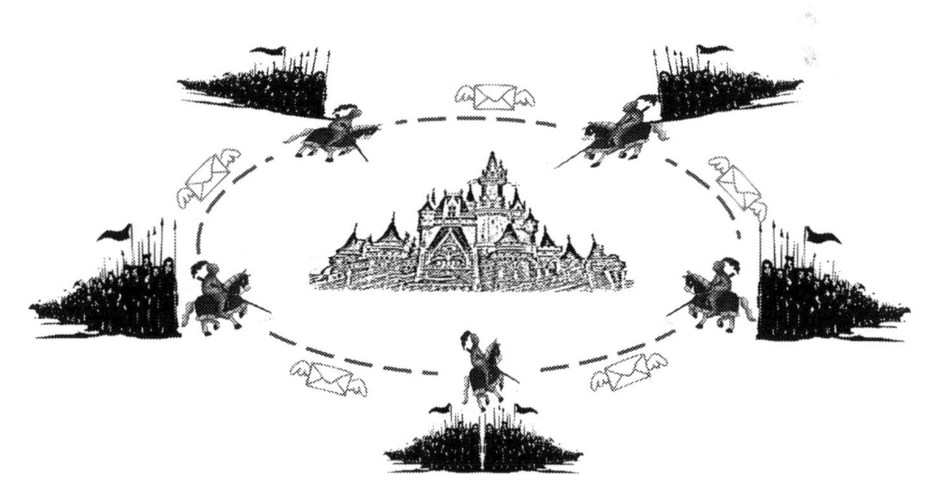

Figure 2-18. *Byzantine army attacking the city*

There are numerous scenarios in real life that are analogous to the Byzantine Generals' Problem. How a group of people reach consensus on some voting agenda or how to maintain the consistent state of a distributed or decentralized database, or maintaining the consistent state of blockchain copies across nodes in a network are a few examples similar to the Byzantine Generals' Problem. Note, however, that the solutions to these different problems could be quite different in different situations. We will look at how Bitcoin solves the Byzantine Generals' Problem later in this book.

Zero-Sum Games

A zero-sum game in game theory is quite straightforward. In such games, one player's gain is equivalent to another player's loss. Example: One wins exactly the same amount as the opponent loses, which means choices by players can neither increase nor decrease the available resources in a given situation.

Poker, Chess, Go, etc. are a few examples of zero-sum games. To generalize even more, the games where only one person wins and the opponent loses, such as tennis, badminton, etc. are also zero-sum games. Many financial instruments such as swaps, forwards, and options can also be described as zero-sum instruments.

In many real-life situations, gains and losses are difficult to quantify. So, zero-sum games are less common compared with non-zero-sum games. Most financial transactions or trades and the stock market are non-zero-sum games. Insurance, however, is a field where a zero-sum game plays an important role. Just think about how the insurance schemes might work. We pay an insurance premium to the insurance companies to guard against some difficult situations such as accidents, hospitalization, death, etc. Thinking that we are insured, we live a peaceful life and we are fairly compensated by the insurance companies when we face such tough situations. There is certainly a financial backup that helps us survive.

Note that everyone who pays the premium does not meet with accident or get hospitalized, and the ones who do need a lot of money compared with the premium they pay. You see, things are quite balanced here, even considering the operational expenses of the insurance company. Again, the insurance company may invest the premium we pay and get some return on that. Still, this is a zero-sum game.

Just to give you a different example, if there is one open position for which an interview drive is happening, then the candidate who qualifies actually does it at the cost of others' disqualification. This is also a zero-sum game.

You may ask if there is any use in studying about zero-sum games. Just being aware of a zero-sum situation is quite useful in understanding and devising a strategy for any complex problem. We can analyze if we can practically gain in a given situation in which the transactions are taking place.

Why to Study Game Theory

Game theory is a revolutionary interdisciplinary phenomenon bringing together psychology, economics, mathematics, philosophy, and an extensive mix of various other academic areas.

We say that game theory is related to real-world problems. However, the problems are limitless. Are the game theoretic concepts limitless as well? Certainly! We use game theory every day, knowingly or unknowingly, because we always use our brains to take the best strategic action, given a situation. Don't we? If that is so, why study game theory?

Well, there are numerous examples in game theory that help us think differently. There are some theories developed such as Nash Equilibrium that relate to many real-life situations. In many real-world situations, the participants or the players are faced with a decision matrix similar to that of a "prisoner's dilemma." So, learning these concepts not only helps us formulate the problems in a more mathematical way, but also enables us to make the best move. It lets us identify aspects that each participant should consider before choosing a strategic action in any given interaction. It tells us to identify the type of game first; who are the players, what are their objectives or goals, what could be their actions, etc., to be able to take the best action. Much decision-making in real life involves different parties; game theory provides the basis for rational decision-making.

The Byzantine Generals' Problem that we studied in the previou section is widely used in distributed storage solutions and data centers to maintain data consistency across computing nodes.

Computer Science Engineering

As mentioned already, it is clever engineering with the concepts from computer science that stitches the components of cryptography, game theory, and many others to build a blockchain. In this section, we will learn some of the important computer science components that are used in blockchain.

The Blockchain

As we will see, a blockchain is actually a blockchain data structure; in the sense that it is a chain of blocks linked together. When we say a block, it can mean just a single transaction or multiple transactions clubbed together. We will start our discussion with hash pointers, which is the basic building block of blockchain data structure.

A hash pointer is a cryptographic hash pointing to a data block, where the hash pointer is the hash of the data block itself (Figure 2-19). Unlike linked lists that point to the next block so you can get to it, hash pointers point to the previous data block and provide a way to verify that the data has not been tampered with.

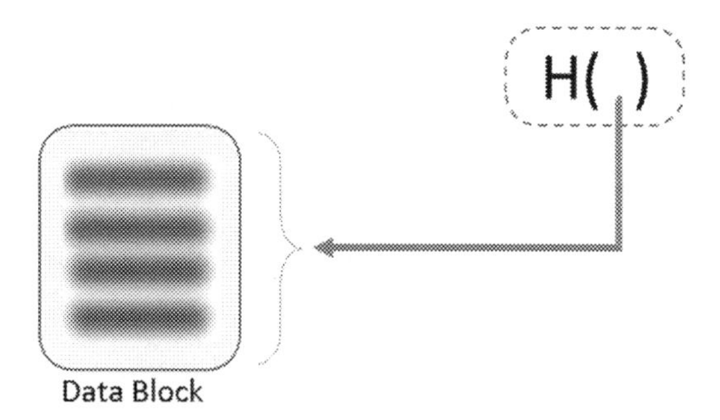

Figure 2-19. *Hash pointer for a block of transactions*

The purpose of the hash pointer is to build a tamper resistant blockchain that can be considered as a single source of truth. How does blockchain achieve this objective? The way it works is that the hash of the previous block is stored in the current block header, and the hash of the current block with its block header will be stored in the next block's header. This creates the blockchain as we can see in Figure 2-20.

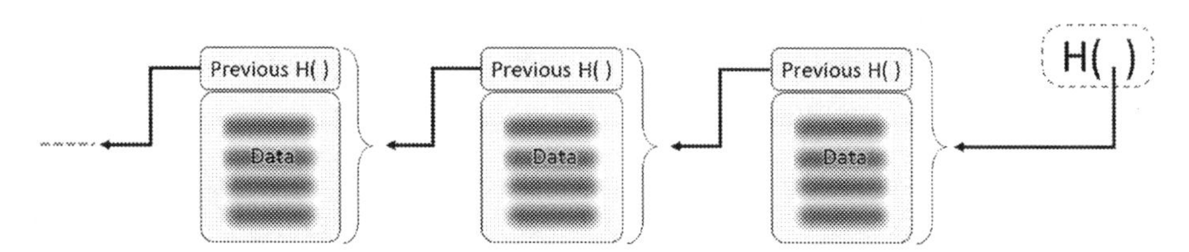

Figure 2-20. *Blocks in a blockchain linked through hash pointers*

As we can observe, every block points to its previous block, known as "the parent block." Every new block that gets added to the chain becomes the parent block for the next block to be added. It goes all the way to the first block that gets created in the blockchain, which is called "the genesis block." In such a design where blocks are linked back with hashes, it is practically infeasible for someone to alter data in any block. We already looked at the properties of hash functions, so we understand that the hashes will not match if the data is altered. What if someone changes the hash as well? Let us focus on Figure 2-21 to understand how it is not possible to alter the data in any way.

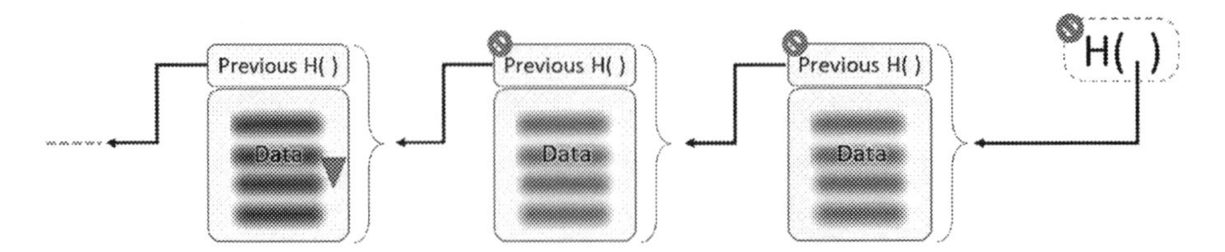

Figure 2-21. *Any attempt in changing Header or Block content breaks the entire chain. Assume that you altered the data in block-1234. If you do so, the hash that is stored in the block header of block-1235 would not match.*

- What if you also change the hash stored in the block header of block-1235 so that it perfectly matches the altered data. In other words, you hash the data block-1234 after you alter it and replace that new hash with the one stored in block header of block-1235. After you do this, the hash of the block-1235 changes (because block-1235 means the data and the header together) and it does not match with the one stored in the block header of block-1236.
- One has to keep doing this all the way till the final or the most recent hash. Since everyone or many in the network already have a copy of the blockchain along with the most recent hash, in no way is it possible to hack into the majority of the systems and change all the hashes at a time.
- This makes it a tamper-proof blockchain data structure.

This clearly means that each block can be uniquely identified by its hash. To calculate this hash, you can use either the SHA2 or SHA3 family of hash functions that we discussed in the cryptography section. If you use SHA-256 to hash the blocks, it would produce a 256-bit hash output such as:

```
000000000000000a73b6a2af7bad40ec3fc2a83dafd76ef15f3d1b71a7132765
```

Notice that there are only 64 characters in it. Since the hashed output is represented using hexadecimal characters, and every hex digit can be represented using four bits, the output is $64 \times 4 = 256$ bits. You would usually see that the 256-bit hashed output is represented using the 64 hex characters in many places.

The structure of a block, that is, block size, the data and header sections, number of transactions in a block, etc., is something that you should decide while designing a blockchain solution. For existing blockchains such as Bitcoin, Ethereum, or Hyperledger, the structure is already defined and you have to understand that to build on top of these platforms. We will take a closer look at the Bitcoin and Ethereum blockchains later in this book.

Merkle Trees

A Merkle tree is a binary tree of cryptographic hash pointers, hence it is a binary hash tree. It is named so after its inventor Ralph Merkle. It is another useful data structure being used in blockchain solutions such as Bitcoin. Merkle trees are constructed by hashing paired data (usually transactions at the leaf level), then again hashing the hashed outputs all the way up to the root node, called the Merkle root. Like any other tree, it is constructed bottom-up. In Bitcoin, the leaves are always transactions of a single block in a blockchain. We will discuss in a little while the advantages of using Merkle trees, so you can decide for yourself if the leaves would be transactions or a group of transactions in blocks. A typical Merkle tree can be represented as in Figure 2-22.

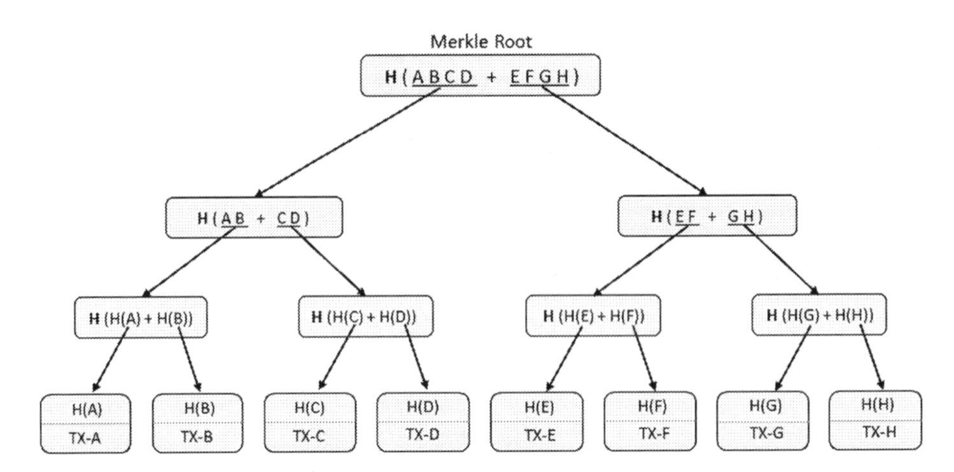

Figure 2-22. *Merkle tree representation*

Similar to the hash pointer data structure, the Merkle tree is also tamper-proof. Tampering at any level in the tree would not match with the hash stored at one level up in the hierarchy, and also till the root node. It is really difficult for an adversary to change all the hashes in the entire tree. It also ensures the integrity of the order of transactions. If you change just the order of the transactions, then also the hashes in the tree till the Merkle root will change.

Here is a situation. The Merkle tree is a binary tree and there should be an even number of items at the leaf level. What if there are an odd number of items? One good solution would be to duplicate the last transaction hash. Since it is the hash we are duplicating, it would mean just the same transaction and not create any issue such as double-spend or repeated transactions. That way, it is possible to balance the tree.

In the blockchain we discussed, if we were to find a transaction through its hash, or check if a transaction had happened in the past, how would we get to that transaction? The only way is to keep traversing till you encounter the exact block that matches the hash of the transaction. This is a case where a Merkle tree can help a great deal.

Merkle trees provide a very efficient way to verify if a specific transaction belongs to a particular block. If there are "n" transactions in a Merkle tree (leaf items), then this verification takes just Log (n) time as shown in Figure 2-23.

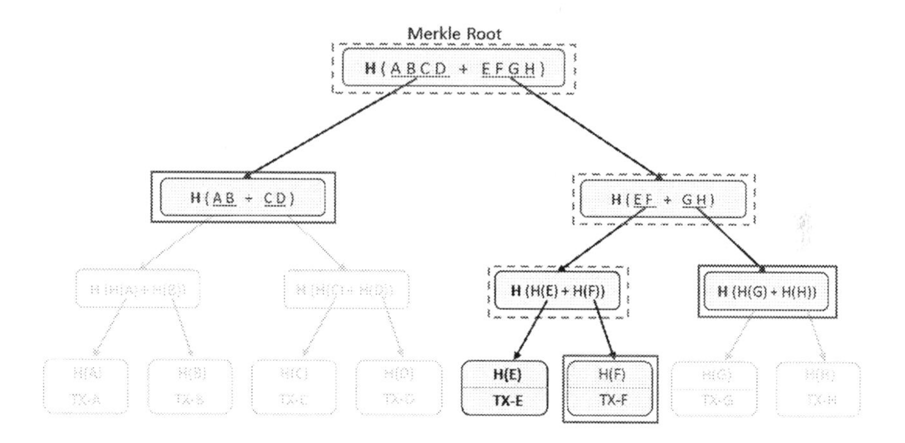

Figure 2-23. *Verification in Merkle tree*

To verify if a transaction or any other leaf item belongs to a Merkle tree, we do not need all items and the whole tree. Rather, a subset of it is needed as we can see in the diagram in Figure 2-23. One can just start with the transaction to verify along with its sibling (it is a binary tree so there would be one sibling leaf item), calculate the hash of those two, and see if it matches their parent hash. Then continue with that parent hash and its sibling at that level and hash them together to get their parent hash. Continuing this process all the way to the top root hash is the quickest possible way for transaction verification (just Log (n) time for n items). In the figure, only the solid rectangles are required and the dotted rectangles can be just computed, provided the solid rectangle data. Since there are eight transaction elements (n = 8), only three computations (log2 8 = 3) would be required for verification.

Now, how about a hybrid of both blockchain data structure and Merkle tree? Imagine a situation in a blockchain where each block has a lot of transactions. Since it is a blockchain, the hash of the previous block is already there; now, including the Merkle root of all the transactions in a block can help in quicker verification of the transactions. If we have to verify a transaction that is claimed to be from, say, block-22456, we can get the transactions of that block, verify the Merkle tree, and confirm quickly if that transaction is valid. We already saw that verifying a transaction is quite easy and fast with Merkle trees. Though blocks in the blockchain are tamper resistant and do not provide even the slightest scope to change anything in a block, the Merkle tree also ensures that the order of transactions is preserved.

In a typical blockchain setting, there could be many situations where a node (for simplicity sake, assume any node that does not have the full blockchain data, i.e., a light node) has to verify if a certain transaction took place in the past. There are actually two things that need verification here: transaction as part of the block, and block as part of the blockchain. To do so, a node does not have to download all the transactions of a block, it can simply ask the network for the information pertaining to the hash of the block and the hash of the transaction. The peers in the network who have the relevant information can respond with the Merkle path to that transaction. Well, you might ask how to trust the data that an unknown peer in the network is sharing with you. You already know that the hash functions are one-way. So in no way can an adversarial node forge transactions that would match a given hash value; it is even difficult to do so from transaction level till the Merkle root.

The use of Merkle trees is not limited to just blockchains: they are widely used in many other applications such as BitTorrent, Cassandra–an NoSQL database, Apache Wave, etc.

Example Code Snippet for Merkletree

This section is just intended to give you a heads-up on how to code up a Merkle tree at its most basic level. Code examples are in Python but would be quite similar in different languages; you just have to find the right library functions to use.

```
# -*- coding: utf-8 -*-

from hashlib import sha256

class MerkelTree(object):
    def __init__(self):
        pass

    def chunks(self,transaction,n):
        #This function yeilds "n" number of transaction at time
        for i in range (0, len(transaction),number):
            yield transaction[i:i+2]

    def merkel_tree(self,transactions):
        #Here we will find the merkel tree hash of all
        transactions passed to this fuction
        #Problem is solved using recursion techqiue

        # Given a list of transactions, we concatinate the
        hashes in groups of two and compute
        # the hash of the group, then keep the hash of group.
        We repeat this step till
        # we reach a single hash
        sub_tree=[]
        for i in chunks(transactions,2):
            if len(i)==2:
                hash = sha256(str(i[0]+i[1])).hexdigest()
            else:
```

```
                hash = sha256(str(i[0]+i[0])).hexdigest()
            sub_tree.append(hash)
        # When the sub_tree has only one hash then we reached
        our merkel tree hash.
        #Otherwise, we call this fuction recursively
        if len(sub_tree) == 1:
            return sub_tree[0]
        else:
            return self.merkel_tree(sub_tree)

if __name__=='__main__':
    mk=MerkelTree()
    merkel_hash= mk.merkel_tree(["TX1","TX2","TX3","TX4","TX5",
    "TX6"])
    print merkel_hash
```

Putting It All Together

To get to this section, we covered all the necessary components of blockchain that can help us understand how it really works. After going through them, namely cryptography, game theory, and computer science engineering concepts, we must have developed a notion of how blockchains might work. Though these concepts have been around for ages, no one could ever imagine how the same old stuff can be used to build a transforming technology such as blockchain. Let us have a quick recap of some fundamentals we covered so far, and we will build further understanding on those concepts. So here they are:

- Cryptographic functions are one-way and cannot be inverted. They are deterministic and produce the same output for a given input. Any changes to the input would produce a completely different output when hashed again.

- Using public key cryptography, digital signatures are possible. It helps in verifying the authenticity of the person/entity that has signed. Considering the private key is kept confidential, it is not feasible to forge a signature with someone else's identity. Also, if someone has signed on any document or a transaction, they cannot later deny they did not.
- Using game theoretic principles and best practices, robust systems can be designed that can sustain in most of the odd situations. Systems that can face the Byzantine Generals' Problem need to be handled properly. Our approach to any system design should be such that the participants play by the rules to get the maximum payoff; deviating from the protocol should not really benefit them.
- The blockchain data structure, by using the cryptographic hashes, provides a tamper resistant chain of blocks. The usage of Merkle trees makes the transaction verification easier and faster.

With all these concepts in mind, let us now think of a real blockchain implementation. What problems can you think of that need to be addressed for such a decentralized system to work properly? Well, there are loads of them; some would be generic to most of the blockchain use cases and some would be specific to a few. Let us discuss at least some of the scenarios that need to be addressed:

- Who would maintain the distributed ledger of transactions? Should all the participants maintain, or only a few would do? How about the computing nodes that are not powerful enough to process transactions or do not have enough storage space to accommodate the entire history of transactions?

- How is it possible to maintain a single consistent state of the distributed ledger? Network latency, packet drops, deliberate hacking attempts, etc. are inevitable. How would the system survive all these?
- Who would validate or invalidate the transactions? Would only a few authorized nodes validate, or all the nodes together would reach a consensus? What if some of the nodes are not available at a given time?
- What if some computing nodes deliberately want to subvert the system or try to reject some of the transactions?
- How would you upgrade the system when there is no centralized entity to take the responsibility? In a decentralized network, what if a few computing nodes upgrade themselves and the rest don't?

There are in fact a lot more concerns that need to be addressed apart from the ones just mentioned. For now we will leave you with those thoughts, but most of those queries should be clarified by the end of this chapter.

Let us start with some basic building blocks of a blockchain system that may be required to design any decentralized solution.

Properties of Blockchain Solutions

So far, we have only learned the technical aspects of blockchain solutions to understand how blockchains might work. In this section, we will learn some of the desired properties of blockchains.

Immutability

It is the most desired property to maintain the atomicity of the blockchain transactions. Once a transaction is recorded, it cannot be altered. If the transactions are broadcast to the network, then almost everyone has a copy of it. With time, when more and more blocks are added to the blockchain, the immutability increases and after a certain time, it becomes completely immutable. For someone to alter the data of so many blocks in a series is not practically feasible because they are cryptographically secured. So, any transaction that gets logged remains forever in the system.

Forgery Resistant

A decentralized solution where the transactions are public is prone to different kinds of attacks. Attempts at forgery are the most obvious of all, especially when you are transacting anything of value. Cryptographic hash and digital signatures can be used to ensure the system is forgery resistant. We already learned that it is computationally infeasible to forge someone else's signature. If you make a transaction and sign a hash of it, no one can alter the transaction later and say you signed a different transaction. Also, you cannot later claim you never did the transaction, because it is you who signed it.

Democratic

Any peer-to-peer decentralized system should be democratic by design (may not be fully applicable to the private blockchain, which we will park for later). There should not be any entity in the system that is more powerful than the others. Every participant should have equal rights in any situation, and decisions are made when the majority reaches a consensus.

Double-Spend Resistant

Double-spend attacks are quite common in monetary as well as nonmonetary transactions. In a cryptocurrency setting, a double-spend attempt is when you try to spend the same amount to multiple people. Example: You have $100 in your account and you pay $90 to two or more parties is a type of double-spend. This is a little different when it comes to cryptocurrency such as Bitcoin where there is no notion of a closing balance. Input to a transaction (when you are paying to someone) is the output of another transaction where you have received at least the amount you are paying through this transaction. Assume Bob received $10 from Alice some time back in a transaction. Today if Bob wants to pay Charlie $8, then the transaction in which he received $10 from Alice would be the input to transact with Charlie. So, Bob cannot use the same input (Alice's $10 paid to him) multiple times to pay to other people and double-spend. Just to give you a different example: if someone owns some land and sells the same piece of land to two people.

In a centralized system it is quite easy to prevent double-spend because the central authority is aware of all the transactions. A blockchain solution should also be immune to such double-spend attacks. While cryptography ensures authenticity of a transaction, it cannot help prevent double-spend. Because, technically, both a normal transaction and a double-spend transaction are genuine. So, the only way possible to prevent double-spend is to be aware of all the transactions. If we are aware of all transactions that happened in the past, we can figure out if a transaction is an attempt to double-spend. So, the nodes that would validate the transactions should definitely be accessible to the whole blockchain data since the genesis block.

Consistent State of the Ledger

The properties we just discussed ensure that the ledger is consistent throughout, to some extent. Imagine a situation when some nodes deliberately want a transaction to not go through and to get rejected. Or, if somehow some nodes are not in sync with the ledger and hence not aware of a few transactions that took place while they were offline, then to them a transaction may look like fraudulent. So, how to ensure consensus among the participants is something that needs to be handled very carefully. Recollect the Byzantine Generals' Problem. The right kind of consensus suitable for a given situation plays the most important role to ensure stability of a decentralized solution. We will learn different consensus mechanisms later in this book.

Resilient

The network should be resilient enough to withstand temporary node failures, unavailability of some computing nodes at times, network latency and packet drops, etc.

Auditable

A blockchain is a chain of blocks that are linked together through hashes. Since the transaction blocks are linked back till the genesis block, auditability already exists and we have to ensure that it does not break at any cost. Also, if one wants to verify whether a transaction took place in the past, then such verification should be quicker.

Blockchain Transactions

When we say blockchain, we mean a blockchain of transactions, right? So it starts from a transaction and then the transaction goes through a series of steps and ultimately resides in the blockchain. Since blockchain is a peer-to-peer phenomenon, if you are dealing with a use case that has a lot of transactions taking place every second, you may not want to flood the whole network with all transactions. Obviously when an individual or an entity is making a transaction, they just have to broadcast it to the whole network. Once that happens, it has to be validated by multiple nodes. Upon validation, it has to again get broadcast to the whole network for the transaction to get included in the blockchain. Now, why not a transaction chain instead of a blockchain? It may make sense to some extent if your business case does not involve a lot of transactions. However, if there are a huge number of transactions every second, then hashing them at transaction level, keeping a trail of it, and broadcasting that to the network can make the system unstable. You may want a certain number of transactions to be grouped in a block and broadcast that block. Broadcasting individual transactions can become a costly affair. Another good reason for a blockchain instead of a transaction chain is to prevent Sybil Attack. In Chapter 3, you will learn in more detail how the PoW mining algorithm is used and one node is chosen at random that could propose a block. If it was not the case, people might create replicas of their own node to subvert the system.

In its most simplified form, the blockchain transactions go through the following steps to get into the blockchain:

- Every new transaction gets broadcast to the network so that all the computing nodes are aware of that fact at the time it took place (to ensure the system is double-spend resistant) .
- Transactions may get validated by the nodes to accept or reject by checking the authenticity.
- The nodes may then group multiple transactions into blocks to share with the other nodes in the network.

- Here comes the difficult situation. Who would propose the block of transactions that they have grouped individually? Broadly speaking, the generation of new blocks should be controlled but not in a centralized fashion, and the mechanism should be such that every node is given equal priority. Every node agreeing upon a block is called the consensus, but there are different algorithms to achieve the same objective, depending on your use case. We will discuss different consensus mechanisms in the following section.
- Though there is no notion of a global time due to network latency, packet drops, and geographic locations, such a system still works because the blocks are added one after another in an order. So, we can consider that the blocks are time stamped in the order they arrive and get added in the blockchain.
- Once the nodes in the network unanimously accept a block, then that block gets into the blockchain and it includes the hash of the block that was created right before it. So this extends the blockchain by one block.

We already discussed the blockchain data structure and the Merkle trees, so we understand their value now. Recollect that when a node would like to validate a transaction, it can do so more efficiently by the Merkle path. The other nodes in the network do not have to share the full block of data to justify proof of membership of a transaction in a block. Technically speaking, memory efficient and computer-friendly data structures such as "Bloom filters" are widely used in such scenarios to test the membership.

Also, note that for a node to be able to validate a transaction, it should ideally have the whole blockchain data (transactions along with their metadata) locally. You should select an efficient storage mechanism that the nodes will adopt based on your use case.

Distributed Consensus Mechanisms

When the nodes are aware of the entire history of transactions by having a local copy of the full blockchain data to prevent double-spend, and they can verify the authenticity of a transaction through digital signatures, what is the use of consensus? Imagine the presence of one or more malicious nodes. Can't they say an invalid transaction is a valid one, or vice versa? Recollect the Byzantine Generals' Problem, which is most likely to occur in many decentralized systems. To overcome such issues, we need a proper consensus mechanism in place.

So far in our discussion, the one thing that is not clear yet is who proposes the block. Obviously, not every node should propose a block to the rest of the nodes at the same time because it is only going to create a mess; forget about the consistent state of the ledger. On the other hand, had it been the case with just transactions without grouping them into blocks, you could argue that if every transaction gets broadcast to the whole network and every node in the network casts a vote on those individual transactions, it would only complicate the system and lead to poor performance.

So, grouping transactions into blocks is important for obvious reasons and consensus is required on a block by block basis. The best strategy for this problem is that only one block should propose a block at a time and the rest of the nodes should validate the transactions in the block and add to their blockchains if transactions are valid. We know that every node maintains its own copy of the ledger and there is no centralized source to sync from. So, if any one node proposes a block and the rest of the nodes agree on it, then all those nodes add that block to their respective blockchains. In such a design, you would prefer that there are at least a few minutes of gap in block creation and it should not be the case where multiple blocks arrive at the same time. Now the question is: who might be that lucky node to propose a block? This is the trickiest part and can lead to proper consensus; we will discuss this aspect under different consensus mechanisms.

These consensus mechanisms actually come from game theory. Your system should be designed such that the nodes get the most benefit if they play by the rules. One of the aspects to ensure the nodes behave honestly is to reward for honest behavior and punish for fraudulent activities. However, there is a catch here. In a public blockchain such as Bitcoin, one can have many different public identities and they are quite anonymous. It gets really difficult to punish such identities because they have a choice to avoid that punishment by creating new identities for themselves. On the other hand, rewarding them works great, because even if someone has multiple identities, they can happily reap the rewards given to them. So, it depends on your business case: if the identities are anonymous, then punishing them may not work, but may work well if the identities are not anonymous. You may want to consider this reward/punish aspect despite having a great mechanism to select a node that would propose the next block. This is because you would never know in advance if the node selected is a malicious node or an honest one. Keep in mind the term *mining* that we may be using quite often, and it would mean generating new blocks.

The goal of consensus is also to ensure that the network is robust enough to sustain various types of attacks. Irrespective of the types of consensus algorithms one may choose depending on the use case, it has to fall into the Byzantine fault tolerant consensus mold to be able to get accepted. Let us now learn some of the consensus mechanisms pertaining to the blockchain scenarios that we may be able to use in different situations.

Proof of Work

The PoW consensus mechanism has been around for a long time now. However, the way it was used in Bitcoin along with other concepts made it even more popular. We will discuss this consensus mechanism at its basic level and look at how it is implemented in Bitcoin in Chapter 3.

The idea behind the PoW algorithm is that certain work is done for a block of transactions before it gets proposed to the whole network. A PoW is actually a piece of data that is difficult to produce in terms of computation and time, but easy to verify. One of the old usages of PoW was to prevent email spams. If a certain amount of work is to be done before one can send an email, then spamming a lot of people would require a lot of computation to be performed. This can help prevent email spams. Similarly, in blockchain as well, if some amount of compute-intensive work is to be performed before producing a block, then it can help in two ways: one is that it will definitely take some time and the second is, if a node is trying to inject a fraudulent transaction in a block, then rejection of that block by the rest of the nodes will be very costly for the one proposing the block. This is because the computation performed to get the PoW will have no value.

Just think about proposing a block without much of effort vs. doing some hard work to be able to propose a block. If it was with almost no effort, then proposing a node with a fraudulent transaction and getting rejected would not have been a big concern. People may just keep proposing such blocks with a hope that one may get through and make it to the blockchain sometime. On the contrary, doing some hard work to propose a block prevents a node from injecting a fraudulent transaction in a subtle way.

Also, the difficulty of the work should be adjustable so that there is a control over how fast the blocks can get generated. You must be thinking, if we are talking about some work that requires some computation and time, what kind of work must it be? It is very simple yet tricky work. An example would help here. Imagine a problem where you have to find a number which, if you hash, the hashed output would start with the alphabet "a." How would you do it? We have learned about the hash functions and know that there are no shortcuts to it. So you would just keep guessing (maybe take any number and keep incrementing by one) the numbers and keep hashing them to see if that fits the bill. If the difficulty level needs to be increased, then one can say it starts with three consecutive "a"s. Obviously, finding a solution for something like "axxxxxxx" is easier to find compared with "aaaxxxxx" because the latter is more constrained.

In the example just given, if multiple different nodes are working to solve such a computational puzzle, then you will never know which node would solve it first. This can be leveraged to select a random node (this time it is truly random because there is no algorithm behind it) that solves the puzzle and proposes the block. It is extremely important to note that in case of public blockchains, the nodes that are investing their computing resources have to be rewarded for honest behavior, else it would be difficult to sustain such a system.

Proof of Stake

The Proof of Stake (PoS) algorithm is another consensus algorithm that is quite popular for distributed consensus. However, what is tricky about it is that it isn't about mining, but is about validating blocks of transactions. There are no mining rewards due to generation of new coins, there are only transaction fees for the miners (more accurately validators, but we will keep using 'miners' so it gets easier to explain).

In PoS systems, the validators have to bond their stake (mortgage the amount of cryptocurrency thay would like to keep at stake) to be able to participate in validating the transactions. The probability of a validator producing a block is proportional to their stake; the more the amount at stake, the greater is their chance to validate a new block of transactions. A miner only needs to prove they own a certain percentage of all coins available at a certain time in a given currency system. For example, if a miner owns 2% of all Ether (ETH) in the Ethereum network, they would be able to mine 2% of all transactions across Ethereum. Accordingly, who gets to create the new block of transaction is decided, and it varies based on the PoS algorithm you are using. Yes, there are variants of PoS algorithm such as naive PoS, delegated PoS, chain-based PoS, BFT-style PoS, and Casper

PoS, to name a few. Delegated PoS (DPOS) is used by Bitshares and Casper PoS is being developed to be used in Ethereum.

Since the creator of a block in a PoS system is deterministic (based on the amount at stake), it works much faster compared with PoW systems. Also, since there are no block rewards and just transaction fees, all the digital currencies need to be created in the beginning and their total amount is fixed all through.

The PoS systems may provide better protection against malicious attacks because executing an attack would risk the entire amount at stake. Also, since it does not require burning a lot of electricity and consuming CPU cycles, it gets priority over PoW systems where applicable.

PBFT

PBFT is the acronym for the Practical Byzantine Fault Tollerance algorithm, one of the many consensus algorithms that one can consider for their blockchain use case. Out of so many blockchain initiatives, Hyperledger, Stellar, and Ripple are the ones that use PBFT consensus.

PBFT is also an algorithm that is not used to generate mining rewards, similar to PoS algorithms. However, the technicalities in their respective implementations are different. The inner working of PBFT is beyond the scope of this book, but at a high level, requests are broadcast to all participating nodes that have their own replicas or internal states. When nodes receive a request, they perform the computation based on their internal states. The outcome of the computation is then shared with all other nodes in the system. So, every node is aware of what other nodes are computing. Considering their own computation results along with the ones received from ther nodes, they make a decision and commit to a final value, which is again shared across the nodes. At this moment, every node is aware of the final decision of all other nodes. Then they all respond with their final decisions and, based on the majority, the final consensus is achieved. This is demonstrated in Figure 2-24.

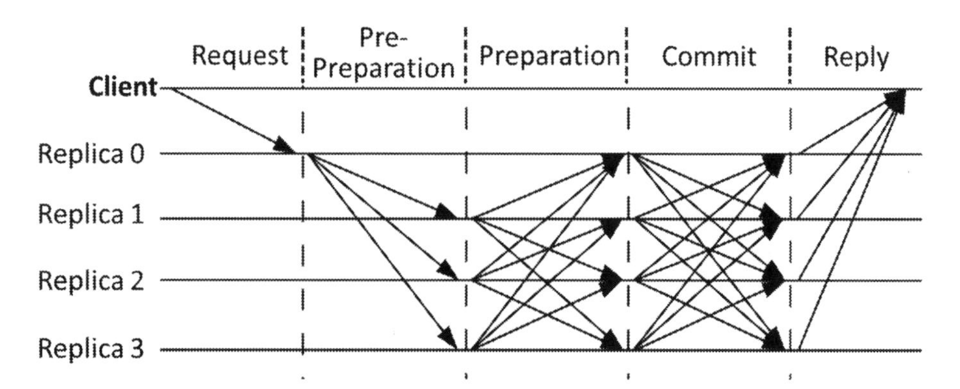

Figure 2-24. *PBFT consensus approach*

PBFT can be efficient compared with other consensus algorithms, based on the effort required. However, anonymity in the system may be compromised because of the way this algorithm is designed. It is one of the most widely used algorithms for consensus even in non-blockchain environments.

Blockchain Applications

While we looked at the nuts and bolts of blockchain throughout this chapter, it is also important that we look at how it is being used in building blockchain solutions. There are applications being built that treat blockchain as a backend database behind a web server, and there are applications that are completely decentralized with no centralized server. Bitcoin blockchain, for example, is a blockchain application where there is no server to send a request to! Every transaction is broadcast to the entire network. However, it is possible that a web application is built and hosted in a centralized web server, and that makes Bitcoin blockchain updates when required. Take a look at Figure 2-25 where a Bitcoin node broadcasts the transactions to the nodes that are reachable at a given point in time.

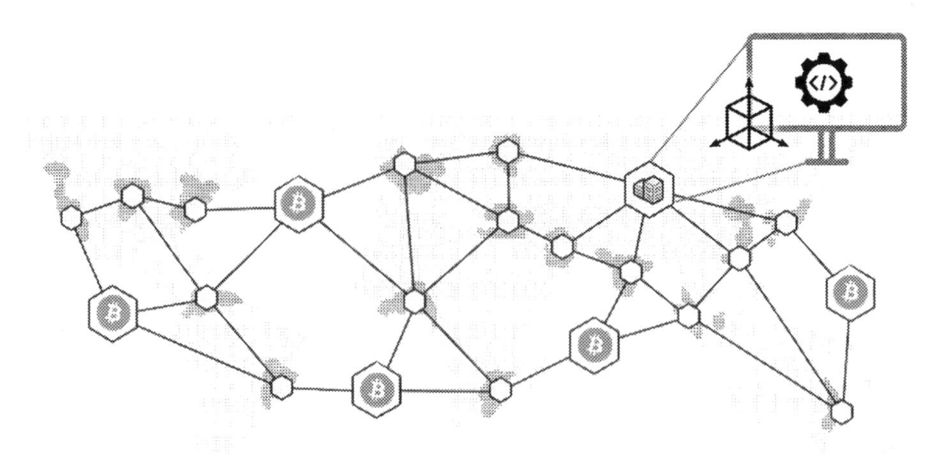

Figure 2-25. *Bitcoin blockchain nodes*

From a software application perspective, every node is self-sufficient and maintains its own copies of the blockchain database. Considering Bitcoin blockchain as a benchmark, the blockchain applications with no centralized servers appear to be the purest decentralized applications and most of them fall under the "public blockchain" category. Usually for such public blockchains, usage of resources from cloud service providers such as Microsoft Azure, IBM Bluemix, etc. are not quite popular yet. For most of the private blockchains, however, the cloud service providers have started to gain popularity. To give you an analogy, there could be one or more web applications for different departments or actors, all of them having their own Blockchin backends and still the blockchains are in sync with each other. In such a setting, though technical decentralization is achieved, politically it could still be centralized. Even though control or governance is enforced, the system is still able to maintain transparency and trust because of the accessibility to single source of truth. Take a look at Figure 2-26, which may resemble most of the blockchain POCs or applications being built on blockchain where blockchains are hosted by some cloud service provider by consuming their blockchain-as-a-Service (BaaS) offering.

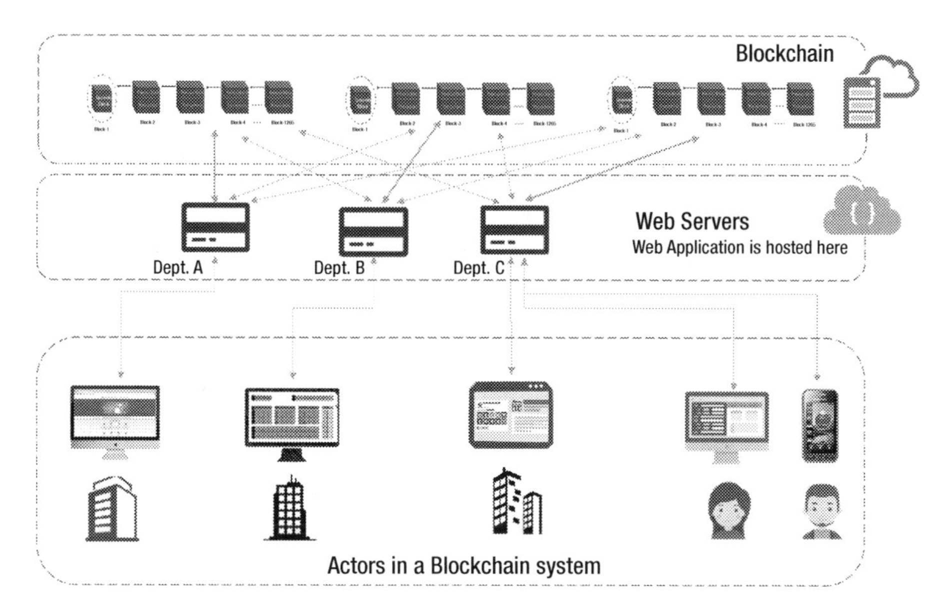

Figure 2-26. *Cloud-powered blockchain system*

It may not be necessary that all the departments have their own different web application. One web application can handle requests from multiple different actors in the system with proper access controm mechanisms. It might be a good idea that all the actors in the system have their own copies of blockchains. Having a local copy of blockchain not only helps maintain transparency in the system, but also may help generate data-driven insights with ready access to data all the time. The different "blockchains" maintained by different actors in the system are consistent by design, thanks to consensus algorithms such as PoW, PoS, etc. Most of the private blockchains prefer any consensus algorithm other than PoW to mitigate heavy resource consumption, and save electricity and computing power as much possible. The PoS consensus mechanism is quite common when it comes to private or consortium blockchains. Since blockchain is disrupting many aspects of businesses, and there was no better way of enabling transparency among them, creating a blockchain solution in the cloud with a "pay as you use" model is gaining momentum. Cloud services are helping businesses leapfrog in their blockchain-enabled digital transformation journey with minimal upfront investments.

There are also decentralized applications (DApps) being built on Ethereum blockchain networks. These applications could be permissioned on private Ethereum or could be permissionless on a public Ethereum network. Also, these applications could be for different use cases on the same public Ethereum network. Though we will cover the Ethereum-specific details later in this book, just look at Figure 2-27 for a high-level understanding of how those applications might look.

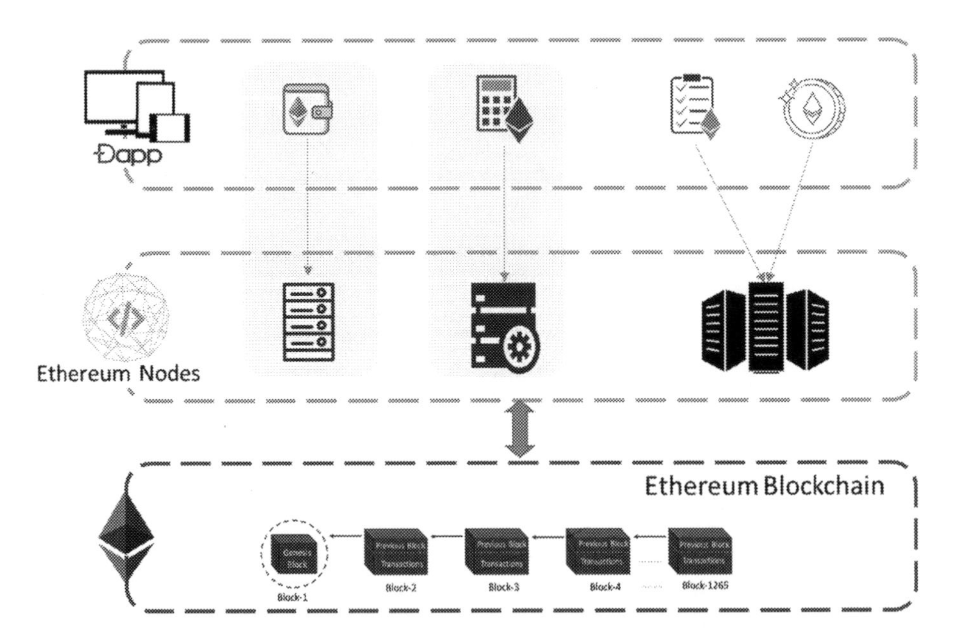

Figure 2-27. *DApps on Ethereum network*

As discussed already in previous sections, developing blockchain applications is only limited by your imagination. Pure blockchain native applications could be built. Applications that treat blockchain as just a backend are also being built, and there are hybrid applications that are also being built that use the legacy applications and use blockchain for some specific purpose only. So far, blockchain scalability is one of the biggest concerns. Though the scalability itself is in research, let us learn some of the scalability techniques.

Scaling Blockchain

We looked at blockchain from a historic perspective and how it proves to be one of the most disruptive technologies as of today. While exploring it technically in this chapter, we learned about the scalability issues inherent to most of the Blockchin flavors. By design, blockchains are difficult to scale and thus a research area in academia and for some innovation-driven corporates. If you look at the Bitcoin adoption, it is not being used to replace fiat currencies due to the inherent scalability challenges. You cannot buy a coffee using Bitcoin and wait for an hour for the transaction to settle. So, Bitcoins are being used as an asset class for investers to invest in. A Bitcoin blockchain network is not capable of accommodating as many transactions as that of Visa or MasterCard, as of today.

Recollect the consensus protocols we have studied so far, such as PoW of Bitcoins or Ethereum, or PoS and other BFT consensus of some other blockchain flavors such as Multichain, Hyperledger, Ripple, or Tendermint. All of these consensus algorithms' primary objective is Byzantine fault tolerance. By design, every node (at least the full nodes) in a blockchain network maintains its own copy of the entire blockchain, validates all transactions and blocks, serves requests from other nodes in the network, etc. to achieve decentralization, which becomes a bottleneck for scalability. Look at the irony here–we add more servers in a centralized system for scalability, but the same does not apply in a decentralized system because with more number of nodes, the latency only increases. While the level of decentralization could increase with a greater number of nodes in a decentralized network, the number of transactions in the network also increases, which leads to increased requirements of computing and storage resources. Keep in mind that this situation is applicable more on public blockchains and less so for private blockchains. Private blockchains could easily scale compared with the public ones because the controlling entities could define and set node specifications with high computation power and more bandwidth. Also, there could be certain tasks offloaded from blockchain and computed off-chain that could help the system scale well.

In this chapter, we will learn some of the generic scaling techniques, and discuss Bitcoin- and Ethereum-specific scaling techniques in their respective chapters. Please keep in mind that all scaling techniques may not apply to all kinds of blockchain flavors or use cases. The best way is to understand the techniques technically and use the best possible one in a given situation.

Off-Chain Computation

Off-chain computation is one of the most promising techniques to scale blockchain solutions. The idea is to limit the usage of blockchain and do the heavy lifting outside of it, and only store the outcomes on blockchain. Keep in mind that there is no standard definition of how the off-chain computation should happen. It is heavily dependent on the situation and the people trying to address it. Also, different blockchain flavors may require different approaches for off-chain computation. At a high level, it is like another layer on top of blockchain that does heavy, compute-intensive work and wisely uses the blockchain. Obviously, you may not be able to retain all the characteristics of blockchain by doing computations off-chain, but it is also true that you may not need blockchain for all kinds of computing requirements and may use it only for specific pain points.

The off-chain computations could be on a sidechain, could be distributed among a random group of nodes, or could be centralized as well. The side chains are independent of the main blockchain. It not only helps scale the blockchain well, it also isolates damages to the sidechain and prevents the main blockchain from any damages from a sidechain. One such example sidechain is the "Lightning Network" for Bitcoins that should help in faster execution of transactions with minimal fee; that will support micropayments as well. Another example of a sidechain for Bitcoins is "Zerocash," whose primary objective is not really scalability, but privacy. If you are using Zerocash for Bitcoin transactions, you cannot be tracked and your privacy is preserved. We will limit our discussion to the generic scalability techniques and not get into a detailed discussion of Bitcoin scalability in this book.

One obvious question that might come up at the moment is how people would check the authenticity of the transactions if they are sent off-chain. First, to create a valid transaction, you do not need a blockchain. We learned in the "Cryptography" section in this chapter about the assymetric key cryptography that is used by the blockchain system. To make a transaction, you have to be the owner of a private key so you can sign the transaction. Once the transaction is created, there are advantages when it gets into the blockchain. Double-spend is not possible with Bitcoin blockchain, and there are other advantages, too. For now, the only objective is to get you on board with the fact that you can create a transaction as long as you own the private key for your account.

Bitcoin blockchains are a stateless blockchain, in the sense that they do not maintain the state of an account. Everything in Bitcoin blockchain is present in the form of a transaction. To be able to make a transaction, you have to consume a previous transaction and there is no notion of "closing balance" for an account, as such. On the contrary, Ethereum blockchain is a "stateful" one! The blocks in Ethereum blockchain contain information regaring the state of the entire block where account balance is also a part. The state information takes up significant space when every node in the network maintains it. This situation is valid for other blockchains as well that are stateful.

Let's take an example to understand this better. Alice and Bob are two parties having multiple transactions between each other. Let's say they usually have 50 monetary transactions in a month. In a stateful blockchain, all these individual transactions would have their state information, and that will be maintained by all the nodes. To address this challenge, the concept of "state channels" is introduced. The idea is to update the blockchain with the final outcome, say, at the end of the month or when a certain transaction threshold is reached, and not with each and every transaction.

State channels are essentially a two-way communication channel between users, objects, or services. This is done with absolute security by using cryptographic techniques. Just to get a heads-up on how it works, take a look at Figure 2-28.

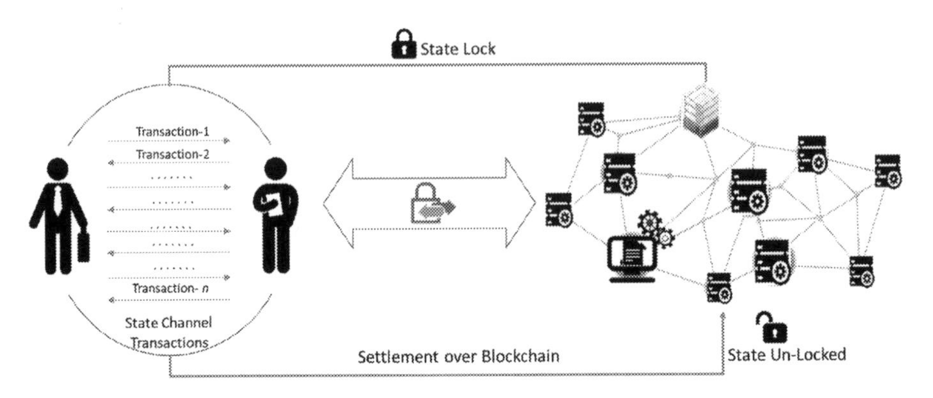

Figure 2-28. *State channels for off-chain computation*

Notice that the off-chain state channels are mostly private and confined among a group of participants. Keep in mind that the state of blockchain for the participants needs to be locked as the first step. Either it could be a MultiSig scheme or a smart contract-based locking. After locking, the participants make transactions among each other that are cryptographically secured. All transactions are cryptographically signed, which makes them verifiable and these transactions are not immediately submitted to the blockchain. As discussed, these state channels could have a predefined lifespan, or could be bound to the amount of transactions being carried out in terms of volume/quantity or any other quantifiable measure. So, the final outcome of the transactions gets settled on the blockchain and that unlocks the state as the final step.

State channels could be very differently implemented in different use cases, and their implementations are actually left to the developers. It is certainly a way forward and is one of the most critical components for mainstream adoption of blockchain applications. For Bitcoin, the Lightning Network was designed for off-chain computation and to make the payments transaction faster. Similarly, the "Raiden Network" was designed for Ethereum blockchain. There are many other such developments to make micropayments faster and more feasible on blockchain networks.

Sharding Blockchain State

Sharding is one of the scalability techniques that has been there for ages and has been a more sought-after topic for databases. People used this technique differently in different use cases to address specific scalability challenges. Before we understand how it could be used in scaling blockchain as well, let us first understand what it means.

Disk read/write has always been a bottleneck when dealing with huge data sets. When the data is partitioned across multiple disks, the read/write could be performed in parallel and latency decreases significantly. This technique is called sharding. Take a look at Figure 2-29.

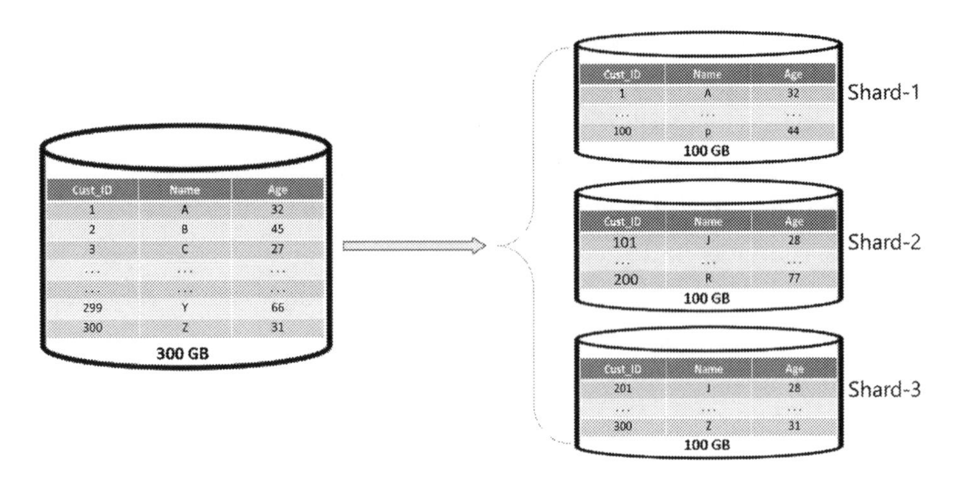

***Figure 2-29.** Database sharding example*

Notice in Figure 2-29 how horizontal partitioning is done to distribute a 300GB database table into three shards of 100GB each and stored on separate server instances. The same concept is also applicable for blockchain, where the overall blockchain state is divided into different shards that contain their own substates. Well, it is definitely not as easy as sharding a database with just doing horizontal partitioning.

So, how does sharding really work in the context of blockchain? The idea is that the nodes wouln't be required to download and keep a copy of the entire blockchain. Instead, they would download and keep the portions (shards) relevant to them. By doing so, they get to process only those transactions that are relevant to the data they store, and parallel execution of transactions is possible. So, when a transaction occurs, it is routed to only specific nodes depending on which shards they affect. If you look at it from a different lens, all the nodes are not required to do all sorts of calculations and verifications for each and every transaction. A mechanism or a protocol could be defined for communication between shards when more than one shard is required to process any specific transactions. Please keep in mind that different blockchains might have different variants of sharding.

To give you an example, you might choose a specific sharding technique for a given situation. One example could be where shards are required to have multiple unique accounts in them. In other words, each unique account is in one shard (more applicable for Ethereum style blockchains that are stateful), and it is very easy for the accounts in one shard to transact among themselves. Obviously, one more level extraction at a shard level is required for sharding to work, and the nodes could keep only a subset of the information.

Summary

In this chapter, we took a deep dive into the core fundamentals of cryptography, game theory, and computer science engineering. The concepts learned would help you design your own blockchain solution that may have some specific needs. Blockchain is definitely not a silver bullet for all sorts of problems. However, for the ones where blockchain is required, it is highly likely that different flavors of blockchain solutions would be needed with different design constructs.

We learned different cryptographic techniques to secure transactions and the usefulness of hash functions. We looked at how game theory could be used to design robust solutions. We also learned some of the core computer science fundamentals such as blockchain data structure and Merkle trees. Some of the concepts were supplimented with example code snippets to give you a jump start on your blockchain assignments.

In the next chapter, we will learn about Bitcoin as a blockchain use case, and how exactly it works.

References

New Directions in Cryptography

Diffie, Whitfield; Hellman, Martin E., "New Directions in Cryptography," IEEE Transactions on Information Theory, Vol IT-22, No 6, https://ee.stanford.edu/~hellman/publications/24.pdf, November, 1976.

Kerckhoff's Principle

Crypto-IT Blog, "Kerckhoff's Principle," www.crypto-it.net/eng/theory/kerckhoffs.html.

Block Cipher, Stream Cipher and Feistel Cipher

http://kodu.ut.ee/~peeter_l/teaching/kryptoi05s/streamkil.pdf.

www.cs.utexas.edu/~byoung/cs361/lecture45.pdf.

www.cs.man.ac.uk/~banach/COMP61411.Info/CourseSlides/Wk2.1.DES.pdf.

https://engineering.purdue.edu/kak/compsec/NewLectures/Lecture3.pdf.

Digital Encryption Standard (DES)

www.facweb.iitkgp.ernet.in/~sourav/DES.pdf.

Advanced Encryption Standard (AES)

www.facweb.iitkgp.ernet.in/~sourav/AES.pdf.

AES Standard Reference

National Institute of Standards and Technology (NIST), "Announcing the Advanced Encryption Standard (AES)," *Federal Information Processing Standards Publication 197,* http://nvlpubs.nist.gov/nistpubs/FIPS/NIST.FIPS.197.pdf, November 26, 2001.

Secured Hash Standard

National Institute of Standards and Technology (NIST), "Announcing the Advanced Encryption Standard (AES)," *Federal Information Processing Standards Publication 197*, http://csrc.nist.gov/publications/fips/fips180-4/fips-180-4.pdf, November 26, 2001.

SHA-3 Standard: Permutation-Based Hash and Extendable-Output Functions

NIST, "Announcing DraftFederl Information Processing Standard (FIPS) 202, SHA-3 Standard: Permutation-Based Hash and Extendable-Output Functions, and Draft Revision of the Applicability Clause of FIPS 180-4, Secure Hash Standard, and Request for Comments," https://csrc.nist.gov/News/2014/Draft-FIPS-202,-SHA-3-Standard-and-Request-for-Com, May 28, 2014.

SHA-3

Paar, Christof, Pelzl, Jan, "SHA-3 and the Hash Function Keccak," *Understanding Cryptography—A Textbook for Students and Practitioners*, (Springer, 2010), https://pdfs.semanticscholar.org/8450/06456ff132a406444fa85aa7b5636266a8d0.pdf.

RSA Algorithm

Kaliski, Burt, "The Mathematics of the RSA Public-Key Cyptosystem," RSA Laboratories, www.mathaware.org/mam/06/Kaliski.pdf.

Milanov, Evgeny, "The RSA Algorithm," https://sites.math.washington.edu/~morrow/336_09/papers/Yevgeny.pdf. June 3, 2009.

Game Theory

Pinkasovitch, Arthur, "Why Is Game Theory Useful in Business?," *Investopedia*, www.investopedia.com/ask/answers/09/game-theory-business.asp, December 19, 2017.

Proof of Stake Algorithm

Buterin, Vitalik, "A Proof of Stake Design Philosophy," *Medium*, https://medium.com/@VitalikButerin/a-proof-of-stake-design-philosophy-506585978d51, December 30, 2016.

Ray, James, "Proof of Stake FAQ," *Ethereum Wiki*, https://github.com/ethereum/wiki/wiki/Proof-of-Stake-FAQ.

Enabling blockchain Innovations with Pegged Sidechains

Back, Adam, Corallo, Matt, Dash Jr, Luke, et al., "Enabling blockchain Innovations with Pegged Sidechains," https://blockstream.com/sidechains.pdf.

CHAPTER 3

How Bitcoin Works

Blockchain technology is all the rage these days, thanks to Bitcoin! blockchain as we know it is a gift of Bitcoin and its inventor, Satoshi Nakamoto, to the whole world. If you are wondering who Satoshi Nakamoto is, it is the name used by the unknown person or persons who originated Bitcoin. We suggest that you understand and appreciate the wonderful technology behind Bitcoin without searching for the inventor. Learning the technical fundamentals of Bitcoin will enable you to understand the other blockchain applications that are there in the market.

Since Bitcoin testified to the robustness of blockchain technology for years, people now believe in it and have started exploring other possible ways to use it. In the previous chapter, we already got the hang of how blockchain works at a technical level, but learning Bitcoin can give you the real taste of blockchain. You may want to consider Bitcoin as a cryptocurrency use case of blockchain technology. So, this chapter will not only help you understand how Bitcoin works in particular, but also give you a perspective of how different use cases can be built using blockchain technology, the way Bitcoin is built.

We will cover Bitcoin in greater detail throughout this chapter and while doing so, a lot of blockchain fundamentals will also be clarified with more practical insights. If you are already familiar with the Bitcoin fundamentals, you may skip this chapter. Otherwise, we advise you to follow through the concepts in the order presented. This chapter explains what Bitcoin is, how it is designed technically, and provides an analysis of some inherent strengths and weaknesses of Bitcoin.

B. Singhal et al., *Beginning Blockchain*, https://doi.org/10.1007/978-1-4842-3444-0_3

The History of Money

Ever wonder what money is and why it even exists? Money is primarily the medium of exchange for exchanging value, that is anything of value. It has a history to it. We will quickly recap the history to be able to understand how money has eveolved to how we know it today, and how Bitcoin furthers it to the next level.

Not everyone has everything. In the good old days when there were no notions of currency or money, people still figured out how they could exchange what they had in surplus for what they needed from someone else. Those were the days of the barter system. Wheat in exchange for peddy or oranges for lemons was the system. This was all good, but what if someone having wheat needs medicine that the other person does not have? Example: Alice has wheat and needs medicine, but Bob knows she has access to someone who has oranges, and Bob needs wheat. In this situation, the exchange is not working out. So, they have to find a third person, Charlie, who might need oranges and as well has surplus medicine. A pictorial representation of this scenario is shown in Figure 3-1.

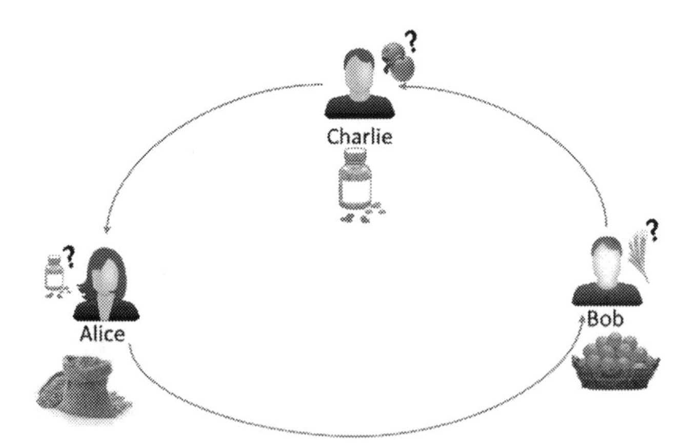

Figure 3-1. *The primitive barter system*

It was always tough to find a person such as Charlie in the previous example who could fit in the puzzle so easily; this problem had to be solved. So, people started thinking of a commoditized system of value exchange. There were a few items that everyone would need, such as milk, salt, seeds, sheep, etc. This system almost worked out! Soon after, people realized that it was quite inconvenient and difficult to store such commodities.

Eventually, better techniques were found to be used as financial instruments, such as metal pieces. People valued the rare metals more than the usual ones. Gold and silver metals topped the list as they wouldn't corrode. Then countries started minting their own currency (metal coins with different weights) with their official seal in them. Though the metal pieces and coins were better than the previous system, as one could easily store and carry them, they were vulnerable to theft. Temples came into rescue as people trusted in them and had a strong belief that no one would steal from temples. The priests would give a receipt to the person depositing gold that would mention the amount of gold/silver received, as a promise to acknowledge their deposit and give back to the bearer of the receipt the same when they returned. The person bearing the receipt could circulate the receipt in the market to get what they wanted. This was the beginning of our banking system. The receipt worked, as the fiat currency and the temples played the role of centralized banks that people trusted. Refer to Figure 3-2 to understand how this system appeared back then.

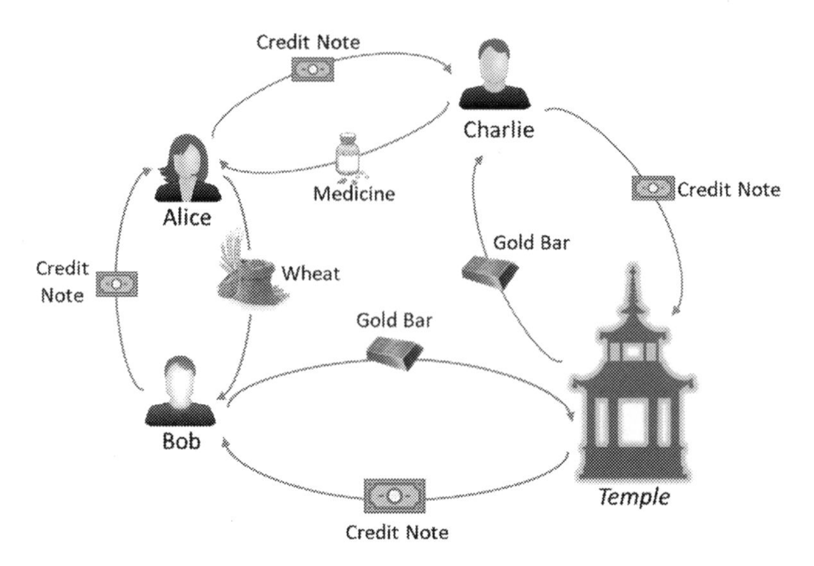

Figure 3-2. *The beginning of the banking era*

In the system just mentioned, currency was always backed by some precious metal such as gold or silver. This system continued even after the goverenments and banks replaced the temples. This is how the commodity currency came up in the market to enable a universal medium of value exchange for the goods and services. Whatever currency was there in those days was all backed by gold/silver.

Slowly, "fiat currency" was introduced by the governments as legal tender, which was no longer backed by gold or silver. It was purely trust based, in the sense that people did not have a choice but to trust the government. Fiat currency does not have any intrinsic value, it is just backed by the government. Today, the money that we know of is all fiat currencies. So, the value of money today depends on the stability and performance of the governments in whose jurisdiction the currency is being issued and used. Those paper currencies were the money themselves and there was nothing more valubale in the banks. This was the state of banking systems and at the same time the digital world was just forming up.

Around the 1990s, the Internet world was gaining momentum and the banking systems were getting digitized. Since some level of discomfort was still there with fiat currencies, since they were perishable and vulnerable to theft, banks assured that people could just go digital! This was the era when even the paper notes weren't required to be printed. Money became the digital numbers in the computer systems of banks. Today, if every account holder went to their respectve bank and demanded the currency notes for the amount of money they hold in their accounts, the banks would be in big trouble! The total real money in circulation is extremely marginal compared with the amount of digital money worldwide.

Dawn of Bitcoin

In the first chapter we looked at the technology aspects of the Internet revolution, and in the previous section of this chapter we looked at the evolution of money. We should now look at them side by side to understand Satoshi Nakamoto's perspective behind designing Bitcoin–a cryptocurrency. In this section and elsewhere in this text, we will try to elaborate on Satoshi's statements in the paper he wrote on Bitcoin.

We learned about temples and then governments and banks for the role they played in the currency systems that eveolved from barter systems. Even today, the situation is just the same. If you zoom in a bit on these systems, you will find that the one pivotal thing that makes these systems stable is the "trust" element. People trusted temples, and then they trusted governments and banks. The entire commerce on the Internet today relies on the centralized, trusted third parties to process payments. Though the Internet was designed to be peer-to-peer, people build centralized systems on it to reflect the same old practice. Well, technically building a peer-to-peer system back in the 2000s was quite tough considering the maturity of technology during that time. Consequently, the cost of transactions, time taken for a transaction to settle, and other issues due to centralization were obvious. This wasn't the case with physical currencies, as transactions meant settlement.

Could there be a digital currency backed by computing power, the same way gold was used to back the money in circulation? The answer is "Yes," thanks to Satoshi's Bitcoins. Bitcoins are designed to enable electronic payments between two parties based on cryptographic proof, and not based on trust due to intermediary third parties. It is possible today because of the technological advancements. In this chapter, we will see how Satoshi Nakamoto combined cryptography, game theory, and computer science engineering fundamentas to design the Bitcoin system in 2008. After it went live in 2009 and till today, the system is quite stable and robust enough to sustain any kind of cyber attacks. It stood the test of time and positioned itself as a global currency.

What Is Bitcoin?

Blockchain offers cryptocurrency: digital money! Just as we can transact with physical currency without banks or other centralized entities, Bitcoin is designed to facilitate peer-to-peer monetary transactions without trusted intermediaries. Let us look at what it is and then learn later in the chapter how it really works. Bitcoin is a decentralized cryptocurrency that is not limited to any nation and is a global currency. It is decentralized in every aspect–technical, logical, as well as political. As and when the transactions are validated, new Bitcoins get mined and a maximum of 21 million Bitcoins can ever be produced. Approximately, to reach 21 million Bitcoins, it would be take until the year 2140. Anyone with good computing power can participate in mining and generate new Bitcoins. After all the Bitcoins get generated, no new coins can be minted and only the ones in circulation would be used. Note that Bitcoins do not have fixed denominations such as the national fiat currencies. As per design, the Bitcoins can have any value with eight decimal places of precision. So, the smallest value in Bitcoin is 0.00000001 BTC, which is called 1 Satoshi.

The miners mine the transactions to mint new coins and also consume the transaction fee that the person willing to make a transaction is ready to pay. When the total number of coins reaches 21 million, the miners would validate the transactions solely for the transaction fees. If someone tries to make a transaction without a transaction fee, it may still get mined because it is a valid transaction (if at all it is) and also the miner is more interested in the mining reward that lets him generate new coins.

Are you wondering what decides the value of Bitcoins? When currency was backed by gold, it had great significance and was easy to assess the value based on gold standards. When we say Bitcoin is backed by the computing power that people use for mining, that is not enough to understand how it attains its value. Here is a litle bit of economics required to understand it.

When fiat currency was launched for the first time, it was backed by gold. Since people believed in gold, they believed in currency as well. After a few decades, currency was no longer backed by gold and was totally dependent on the governments. People continued believing in it because they themselves form or contribute to the formation of their own government. Since the governments ensure its value, and people trust it, so it attains that value. In an international setting, the value of currency of specific countries depends on various factors and the most important of them is "supply and demand." Please keep in mind that some countries that printed a lot of fiat currency notes went bankrupt; their economy went down! There has to be a balance and to understand this, more economics is needed, which is beyond the scope of this book. So, let us get back to Bitcoins for now.

When Bitcoin was first launched, it did not have any official price or value that people would believe. If one would sell it for some US dollars (USD), I would never have bought those initially. Gradually when the exchange started taking place, it developed a price and one Bitcoin was not even one USD then. Since Bitcoins are generated by a competitive and decentralized process called "mining," and they are generated at a fixed rate with an upper cap of 21 million Bitcoins in total that can ever exist, this makes Bitcoin a scarce resource. Now relating this context back to the game of "supply and demand," the value of Bitcoin started inflating. Slowly, when the entire globe started believing in it, its price even skyrocketed from a few USDs to thousands of USDs. Bitcoin adoption among the users, merchants, start-ups, big businesses, and many others is growing like never before because they are being used in the form of money. So, the value of Bitcoin is highly influenced by "trust," "adoption," and "supply and demand" and its price is set by the market.

Now, the question is why the value of Bitcoin is so volatile as of this writing and fluctuates quite a lot. One obvious reason is supply and demand. We learned that there can only be a limited number of Bitcoins in circulation, which is 21 million, and the rate at which they get generated is decreeasing with time. Because of this design, there is always a gap in supply and demand, which results in this volatility. Another reason is that Bitcoins are never traded in one place. There are so many exchanges in so many places across the globe, and all those exchanges have their own exchange prices. The indexes that you see gather Bitcoin exchange prices from several exchanges and then average them out. Again, since all these indexes do not collect data from the same set of exchanges, even they do not match. Similarly, the liquidity factor that implies the amount of Bitcoins flowing through the entire market at any given time also influences the volatility in Bitcoin price. As of now, it is definitely a high-risk asset but may get stabilized with time. Let us take a look at the following list of factors that may influence the supply and demand of Bitcoins, and hence their price:

- Confidence of people in Bitcoin and fear of uncertainty
- Press coverage with good and bad news on Bitcoin

- Some people own Bitcoins and do not allow them to flow through the market and some people keep buying and selling to minimize risk. This is why the liquidity level of Bitcoin keeps on changing.
- Acceptance of Bitcoins by big ecommerce giants
- Banning of Bitcoins in specific countries

If you are now wondering if there is any possibility of Bitcoin to crash completely, then the answer is "Yes." There are many examples of countries whose currency systems have crashed. Well, there were political and economic reasons for them to crash such as hyperinflation, which is not the case with Bitcoins because one cannot generate as many Bitcoins as they want and the total number of Bitcoins is fixed. However, there is a possibility of technical or cryptographic failure of Bitcoins. Please note that Bitcoin has stood the test of time since its inception in 2008 and there is a possibility that it will grow much bigger with time, but it cannot be guaranteed!

Working with Bitcoins

In order to get started with Bitcoins, no technicality is needed. You just have to download a Bitcoin wallet and get started with it. When you download and install a wallet on your laptop or mobile, it generates your first Bitcoin address (public key). You can generate many more, however, and you should. It is a best practice to use the Bitcoin addresses only once. Address reuse is an unintended practice in Bitcoin, though it works. Address reuse can harm privacy and confidentiality. As an example, if you are reusing the same address, signing a transaction with the same private key, the recipient can easily and reliably determine that the address being reused is yours. If the same address is used with multiple transactions, they can all be tracked and finding who you are gets even easier.

Remember that Bitcoin is not fully anonymous; it is said to be pseudonymous and there are ways to trace the transaction origins that can reveal the owners.

You have to disclose your Bitcoin address to the person willing to transfer Bitcoins to you. This is very safe because the public key is public anyway. We know that there is no notion of a closing balance in Bitcoin and all records are there as transactions. Bitcoin wallets can easily calculate their spendable balance, as they have the private keys of the corresponding public keys on which transactions are received. There are a variety of Bitcoin wallets available from so many wallet providers. There are mobile wallets, desktop wallets, browser-based web wallets, hardware wallets, etc., with varying levels of security. You need to be extremely careful in the wallet security aspect while working with Bitcoins. The Bitcoin payments are irreversible.

You must be wondering how secured are these wallets. Well, different wallet types have different leves of security and it depends on how you want to use it. Many online wallet services suffered from security breaches. It is always a good practice to enable two-factor authentication whenever applicable. If you are a regular user of Bitcoins, it may be a good idea to use small amounts in your wallets and keep the remainder separately in a safe environment. An offline wallet or cold wallet that is not connected to the network provides the highest level of security for savings. Also, there should be proper backup mechanisms for your wallet in case you lose your computer/mobile. Remember that if you lose your private key, you lose all the money associated with it.

If you have not joined Bitcoin as a miner running a full node, then you can just be a user or a trader of Bitcoins. You will definitely need an exchange from where you can buy some Bitcoins with your US dollars or other currencies as accepted by the exchanges. You should prefer buying Bitcoins from a legitimate and secured exchange. There have been many examples of exchanges that suffered from security breaches.

The Bitcoin Blockchain

We already looked at the basic blockchain data structure in the previous chapter and also covered the basic building blocks of a blockchain data structure such as hashing techniques and asymmetric cryptography. We will learn the specifics of Bitcoin blockchain in this section.

The Bitcoin blockchain, like any other blockchain, has a similar blockchain data structure. The Bitcoin Core client uses the LevelDB database of Google to store the blockchain datastructure internally. Each block is identified by its hash (Bitcoin uses the SHA256 hashing algorithm). Every block contains the hash of the previous block in its header section. Remember that this hash is not just the hash of the previous header but the entire block including header, and it continues all the way to the *genesis* block. The genesis block is the beginning of any blockchain. Typically, a Bitcoin blockchain looks as shown in Figure 3-3.

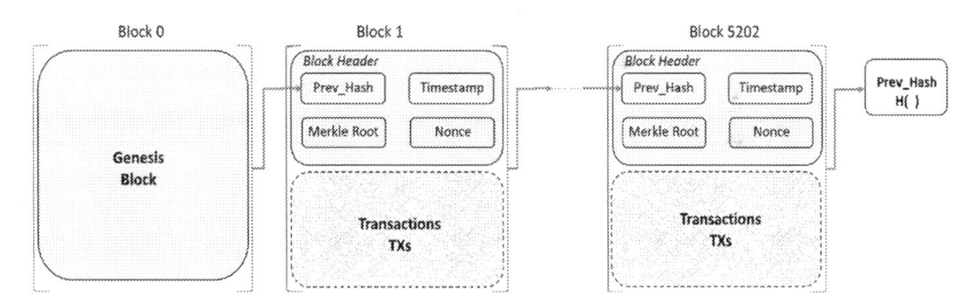

Figure 3-3. *The Bitcoin blockchain*

As you can see in this blockchain, there is a block header part that contains the header information and there is a body part where the transactions are bundled in every block. Every block's header contains the hash of the previous block. So, any change in any block in the chain will not be so easy; all the subsequent blocks have to be changed accordingly. Example: If someone tries to change a previous transaction that was captured in, say, block number 441, after changing the transaction, the hash of this block that is in the header of block number 442 will not match, so it has to be changed as well. Changing the header with the new hash will then require you to update the hash in the block header of the next block in the sequence, which is block number 443, and this will go on till the current block and this is tough work to do. It beccomes almost impossible when we know that every node has it's own copy and hacking into all the nodes, or at least 51% of them, is infeasible.

In the blockchain, there is only one true path to the genesis block. However, if you start from the genesis block, then there can be forks. When two blocks are proposed at the same time and both of them are valid, only one of them would become a part of the true chain and the other gets orphaned. Every node builds on the longest chain, and whichever it hears first and whichever becomes the longest chain will be the one to build on. Such a scenario can be represented as shown in Figure 3-4.

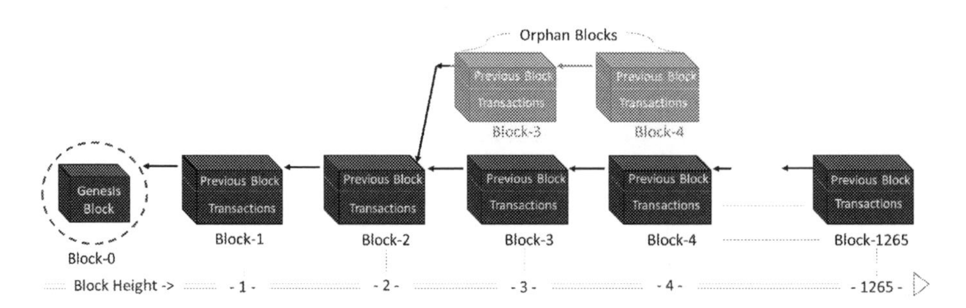

Figure 3-4. *Orphan blocks in true blockchain*

Observe in Figure 3-4 that at block height-3, two blocks are proposed to become block-3, but only one of them could make it to the final blockchain and the rest got orphaned out. It is evident that at a certain block height, there is a possibility of one or more blocks because there can as well be some orphaned blocks at this height, so block height is not the best way to uniquely identify a block and block hash is the right way to do so.

Block Structure

The block structure of a Bitcoin blockchain is fixed for all blocks and has specific fields with their corresponding required data. Take a look at Figure 3-5, a birds-eye view of the entire block structure, and then we will learn more about the individual fields later in this chapter.

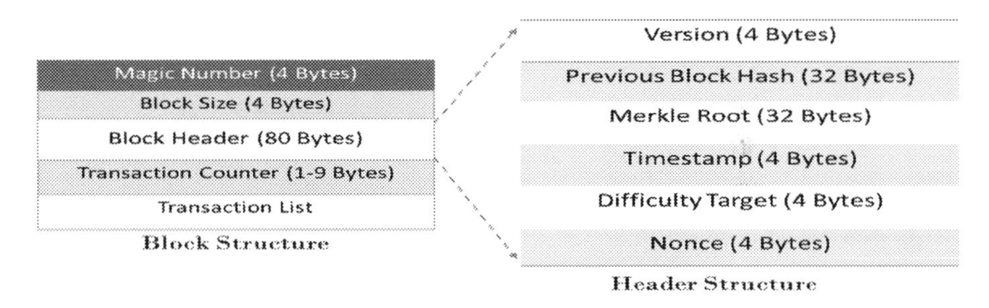

Figure 3-5. *Block structure of Bitcoin blockchain*

A typical block structure appears as shown in Table 3-1.

Table 3-1. *Block Structure*

Field	Size	Description
Magic Number	4 bytes	It has a fixed value 0xD9B4BEF9, which indicates the start of the block and also that the block is from the mainnet or the production network.
Block Size	4 bytes	This indicates the size of the block. The original Bitcoin blocks are of 1MB and there is a newer version of Bitcoin called "Bitcoin Cash" whose block size is 2MB.
Block Header	80 bytes	It comprises much information such as Previous Block's hash, Nonce, Merkle Root, and many more.

(*continued*)

Table 3-1. (*continued*)

Field	Size	Description
Transaction Counter	1–9 bytes (variable length)	It indicates total number of transactions that are included within the block. Not every transaction is of the same size, and there is a variable number of transactions in every block.
Transaction List	Variable in number but fixed in size	It lists all the transactions that are taking place in a given block. Depending on block size (whether 1MB or 2MB), this field occupies the remaining space in a block.

Let us now zoom in (Table 3-2) to the "Block Header" section of the blocks and learn the various different fields that it maintains.

Table 3-2. *Block Header Components*

Field	Size	Description
Version	4 bytes	It indicates the version number of Bitcoin protocol. Ideally each node running Bitcoin protocol should have the same version number.
Previous Block Hash	32 bytes	It contains the hash of the block header of the previous block in the chain. When all the fields in the previous block header are combined and hashed with SHA256 algorithm, it produces a 256-bit result, which is 32 bytes.

(*continued*)

Table 3-2. (*continued*)

Field	Size	Description
Merkle Root	32 bytes	Hashes of the transactions in a block form a Merkle tree by design, and Merkle root is the root hash of this Merkle tree. If a transaction is modified in the block, then it won't match with the Merkle root when computed. This way it ensures that keeping the hash of the previous block's header is enough to maintain the secured blockchain. Also, Merkle trees help determine if a transaction was a part of the block in O(*n*) time, and are quite fast!
Timestamp	4 bytes	There is no notion of a global time in the Bitcoin network. So, this field indicates an approximate time of block creation in Unix time format.
Difficulty Target	4 bytes	The proof-of-work (PoW) difficulty level that was set for this block when it was mined
Nonce	4 bytes	This is the random number that satisfied the PoW puzzle during mining.

The block fields and their corresponding explanations as presented in the previous tables are good enough to start with, and we will explore more of only a few fields that require a more detailed explanation.

Merkle Tree

We have covered the concept of Merkle trees in the previous chapter. In this section, we will just take a look at how Bitcoin uses Merkle trees. Each block in a Bitcoin blockchain contains the hash of all the transactions, and the Merkle root of all these transactions is included in the header of that block. In a true sense, when we say that each block header contains the hash of the entire previous block, the trust is that it just contains the hash of the previous block's header. Nonetheless, it is enough, because the header already contains the Merkle root. If a transaction in the block is altered, the Merkle root will not match anymore and such a design still preserves the integrity of the blockchain.

The Merkle tree is a tree data structure of the hash of the transactions. The "Leaf Nodes" in the Merkle tree actually represent the hash of the transactions, whereas the root of the tree is the Merkle root. Refer to Figure 3-6.

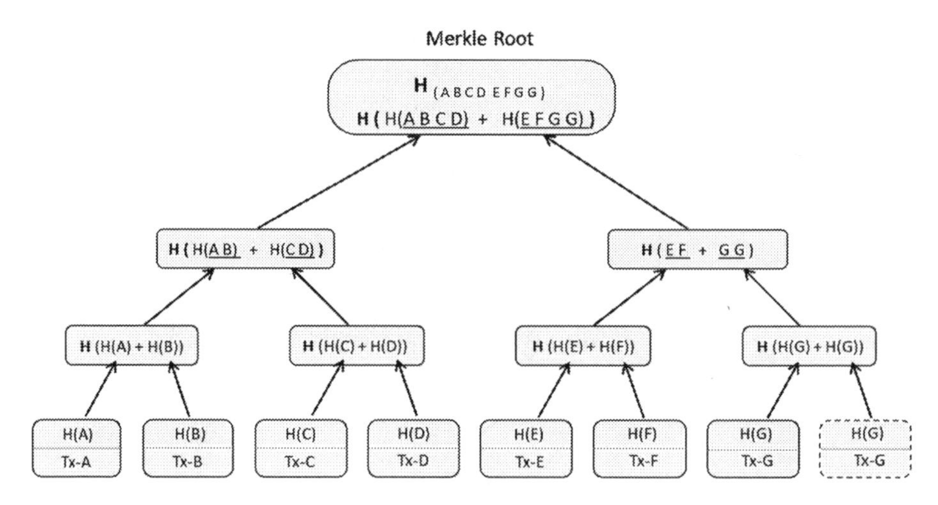

Figure 3-6. *Merkle-tree representation*

Notice that the hash of the seven transactions A, B, C, D, E, F, and G form the leaf of the tree. Since there are seven transactions but the total leaf nodes should be even in a binary tree, the last leaf node gets repeated. Each transaction hash of 32 bytes (i.e., 256 bits) is calculated by applying SHA256 twice to the transactions. Similarly, the hash of two transactions are concatenated (62 bytes) and then hashed twice with SHA256 to get the parent hash of 32 bytes.

Only the Merkle path to a transaction is enough to verify if a transaction was a part of any block and is quite efficient therefore. So, the actual blockchain can be represented as shown in Figure 3-7.

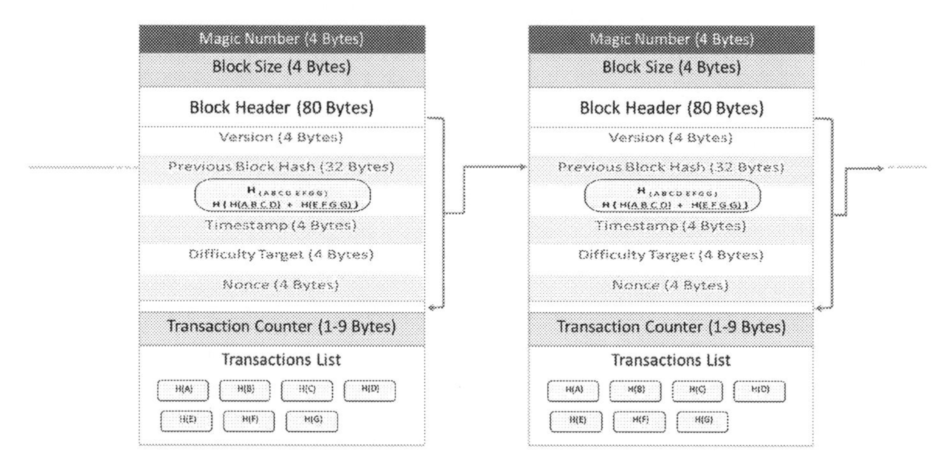

Figure 3-7. *Merkle tree representation*

Difficulty Target

The difficulty target is the one that drives the PoW in Bitcoin. The idea is that once a block is filled with valid transactions, the hash of that block's header needs to be calculated to be less than the difficulty target in the same header. The nonce in the header starts from zero. The miner has to keep on incrementing this nonce and hashing the header till the hash value is less than the target.

The difficulty bits of four bytes (32 bits) in the headers define what would be the *target* value (256 bits) for that block to be mined. The nonce should be found such that the hash of the entire header should be less than the target value. Remember that the lower the target value, the more difficult it would be to find a header hash that would be less than the target. Since Bitcoin uses SHA256, every time you hash a block header, the output is any number between 0 and 2^{256}, which is quite a big number.

If with your nonce the hash is less than the target, the block will be accepted by the entire network, else you have to try with a different nonce till it satisfies the condition. At this point, it is still not clear how the difficulty target is calculated with the difficulty bits in each header.

The target can be derived from the four-byte (8 hexadecimal numbers) difficulty bits in the header using a predefined formula that every node has by default, as it came along with the binaries during installation. Following is the formula to compute the difficulty:

```
target = coefficient * 2^(8 * (exponent - 3))
```

Notice that there is a "coefficient" and there is also an "exponent" term in this formula, which are present as a part of the four-byte difficulty bits. Let us take an example to explain this better. If the four-byte difficulty bits in hex form are 0x1b0404cb, then the first two hex digits form the exponent term, which is (0x1b), and the remaining form the coefficient term, which is (0x0404cb) in this case. Solving for the target formula with these values:

```
target = 0x0404cb * 2^(0x08 * (0x1b - 0x03))
target = 0x00000000000404
CB000000000000000000000000000000000000000000000000
```

Bitcoin is designed such that every 2,016 blocks should take two weeks to be generated and if you do the math, it would be around ten minutes for every block. In this asynchronous network, it is difficult to program like this where every block takes exactly ten minutes with the kind of PoW mechanism in place. In reality, it is the average time for a block, and there is a possibility that a Bitcoin block gets generated within a minute or it may very well take 15 minutes to be generated. So, the difficulty is designed to increase or decrease depending on whether it took less or more than two weeks to find 2,016 blocks. This time taken for 2,016 blocks can be found using the time present in the timestamp fields of every block header.
If it took, say, **T** amount of time for 2,016 blocks, which is never exactly

two weeks, the difficulty target in every block is multiplied by (**T / 2 weeks**). So, the result of [difficulty target × (**T / 2 weeks**)] will be increased if **T** was less and decreased otherwise.

It is evident by now that the difficulty target is adjustable; it could be set more difficult or easier depending on the situation we explained before. You must be wondering, who adjusts this difficulty when the system is decentralized? One rule of thumb that you must always keep in mind is that whatever happens in such a decentralized design has to happen individually at every node. After every 2,016 blocks, all the nodes individually compute the new difficulty target value and they all conclude on the same one because there is already a formula defined for it. To have this formula handy, here it is once again:

```
New Target= Old Target * (T / 2 weeks)
⇨ New Target= Old Target * (Time taken for 2016 Blocks in
Seconds / 12,09,600 seconds)
```

Note The parameters such as 2,016 blocks and TargetTimespan of two weeks (12,09,600 seconds) are defined in **chainparams.cpp** as shown following:

```
consensus.nPowTargetTimespan = 14 * 24 * 60 * 60; // two weeks
consensus.nPowTargetSpacing = 10 * 60;
consensus.nMinerConfirmationWindow = 2016; //
nPowTargetTimespan / nPowTargetSpacing
```

Note here that it is (T / 2 weeks) and not (2 weeks / T). The idea is to decrease the difficulty target when it is required to increase the complexity, so it takes more time. The lesser the target hash, the more difficult it gets to find a hash that is less than this target hash. Example: If it took ten days to mine 2,016 blocks, then (T / 2 weeks) would be a fraction, which when multiplied by "Old Target" further reduces it and "New Target" would be a value less than the old one. This would make it difficult to find a hash and would require more time. This is how the time between blocks is maintained at ten minutes on average. Imagine that the difficulty target was fixed and not adjustable; what do you think the problem would be? Remember that the computation power of the hardware increases with time as more powerful computers are introduced for block mining. A situation where 10s or 100s or even 1,000s of blocks are proposed at the same time is not desirable for the network to function properly. So, the idea is that, even when more and more powerful computing nodes enter into the Bitcoin network, avrage time required to propose a block should still be ten minutes by adjusting the difficulty target. Also, a miner's chances of proposing a block depends on how much hash power they have compared with the global hash power of all miners included.

Are you thinking why ten minutes, and why not 12 minutes? Or why not six minutes? Just keep in mind that there has to be some time gap for all the nodes in a decentralized asynchronous system to agree on it. If there was no time gap, so many blocks would arrive with just fractional delays and there wouldn't be any optimization benefit of **blockchain** as compared with **transaction chain**. Every transaction is a broadcast and every new block is also a broadcast. Also, the orderliness that a blockchain brings to the system is quite infeasible by the transaction chain. With the concept of blocks, it is possible to include the unrelated transactions from any sender to any receiver in blocks, which is not easy to maintain with the transaction chain. One valid block broadcast is more efficient compared with individual transaction broadcast after validation. Now, coming back to the discussion of ten minutes, it can very well be a little less or a little more but there should certainly be some gap between two consecutive blocks. Imagine that you are a miner and mining block number 4567, but some other miner got lucky and proposed block number 4567, which you just received while solving the cryptographic puzzle. What you would do now is validate this block and if it is valid, add it to your local copy of the blockchain and immediately start on mining the 4568. You wouldn't want someone else to propose 4568 already while you just finished validating block 4567, which you received a little later compared with other miners due to network latency. Now the question is: *is this 10 minutes the best possible option*? Well, it is difficult to explain this in one word, but a ten-minute gap addresses a lot of issues due to an asynchronous network, time delays, packet drops, system capacity, and more. There is a possibility that it could be optimized further to, say, five minutes or so, which you can see in many new cryptocurrencies and other blockchain use cases.

The Genesis Block

The very first block as you can see in the following code, the block-0, is called the *genesis* block. Remember that the genesis block has to be hardcoded into the blockchain applications and so is the case with Bitcoin. You can consider it as a special block because it does not contain any reference to the previous blocks. The Bitcoin's genesis block was created in 2009 when it was launched. If you open the Bitcoin Core, specifically the file *chainparams.cpp*, you will see how the genesis block is statically encoded. Using a command line reference to Bitcoin Core, you can get the same information by querying with the hash of the genesis block as shown below:

```
$ bitcoin-cli getblock 000000000019d6689c085ae165831e934
ff763ae46a2a6c172b3f1b60a8ce26f
```

Output of the preceding command:

```
{
    "hash" : "000000000019d6689c085ae165831e934ff763ae46a2a6
    c172b3f1b60a8ce26f",
    "confirmations" : 308321,
    "size" : 285,
    "height" : 0,
```

```
    "version" : 1,
    "merkleroot" : "4a5e1e4baab89f3a32518a88c31bc87f618
    f76673e2cc77ab2127b7afdeda33b",
    "tx" : ["4a5e1e4baab89f3a32518a88c31bc87f618f76673e2cc
    77ab2127b7afdeda33b"],
    "time" : 1231006505,
    "nonce" : 2083236893,
    "bits" : "1d00ffff",
    "difficulty" : 1.00000000,
    "nextblockhash" : "00000000839a8e6886ab5951d76
    f411475428afc90947ee320161bbf18eb6048"
}
```

If you convert the Unix time stamp as shown in the previous output, you will find this date-time information: *Saturday 3rd January 2009 11:45:05 PM.* You can as well get the same information from the website `https://blockchain.info`. Just navigate to this site and paste the hash value in the right top search box and hit "Enter." Here is what you will find (Table 3-3)

Table 3-3. *Transaction Information*

Summary	
Number Of Transactions	1
Output Total	50 BTC
Estimated Transaction Volume	0 BTC
Transaction Fees	0 BTC
Height	0 (Main Chain)
Timestamp	2009-01-03 18:15:05
Received Time	2009-01-03 18:15:05

(*continued*)

Table 3-3. (*continued*)

Summary	
Relayed By	Unknown
Difficulty	1
Bits	486604799
Size	0.285 kB
Weight	0.896 kWU
Version	1
Nonce	2083236893
Block Reward	50 BTC

Table 3-4. *Hash Information*

Hashes	
Hash	000000000019d6689c085ae165831e934ff763ae46a2a6c172b3f1b60a8ce26f
Previous Block	0000000000000000000000000000000000000000000000000000000000000000
Next Block(s)	00000000839a8e6886ab5951d76f411475428afc90947ee320161bbf18eb6048
Merkle Root	4a5e1e4baab89f3a32518a88c31bc87f618f76673e2cc77ab2127b7afdeda33b

In this Block-0, there is just one transaction, which is a coinbase transaction. Coinbase transactions are the ones that the miners get. There are no inputs to such transactions and they can only generate new Bitcoins. If you explored the transactions associated in this block, here is how it would look (Figure 3-8).

Figure 3-8. *Coinbase transaction in Block-0*

The Bitcoin Network

The Bitcoin network is a peer-to-peer network, as discussed already. There is no centralized server in such a system and every node is treated equally. There are no master–slave phenomena and no hierarchy as well in such a system. Since this runs on the Internet itself, it uses the same TCP/IP protocol stack as shown in Figure 3-9.

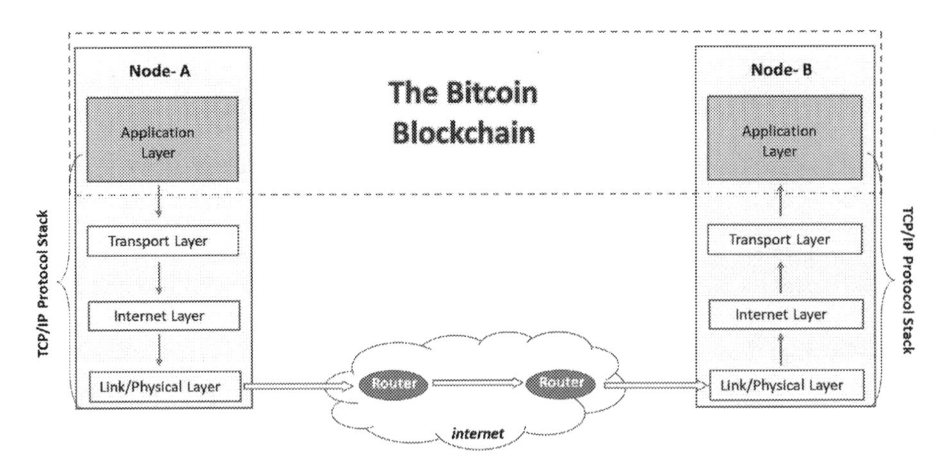

Figure 3-9. *The Bitcoin blockchain network on the Internet*

The above diagram shows how Bitcoin networks coexist on the same Internet stack. The Bitcoin network is quite dynamic in the sense that nodes can join and leave the netwrk at will and the system still works. Also, despite being asynchronous in nature and with network delays and packet drops, the system is very robust–thanks to the design of Bitcoin!

The Bitcoin network is a decentralized network with no central point of failure and as well no central authority. With such a design, how would you assess how big the Bitcoin network is? There is no proper way of estimating this as the nodes can join and leave at will. However, there are some attempts at researching the Bitcoin network, and some claim that there are close to 10,000 nodes that are mostly connected to the network all the time and there can be millions of nodes at a time.

Every node in the Bitcoin network is equal in terms of authority and has a flat structure, but the nodes can be full nodes or lightweight nodes. The full nodes can do almost every permissible activity in the Bitcoin system, such as mining transactions and broadcasting transactions, and can provide wallet services. The full nodes also provide the routing function to participate in and maintain the Bitcoin network. To become a full node, you have to download the entire blockchain database that containss the entire transactions taken place till now. Also, the node must stay permanently connected to the Bitcoin network and hear all transactions taking place. It is important that you have a good network connection, good storage (at least 200GB), and at least 2GB RAM dedicated to it. This requirement may further change and require more resources with time.

On the other hand, lightweight nodes cannot mine new blocks but can verify transactions by using Simplified Payment Verification (SPV). They are otherwse termed "thin clients." A vast majority of nodes in the Bitcoin network are SPVs. They can as well participate in pool mining where there are many nodes trying to mine new blocks together. Lightweight nodes can help verify the transactions for the full nodes. A good example of an SPV is a wallet (the client). If you are running a wallet and someone sends money to you, you can act as a node in the Bitcoin network and download the relevant transactions to the one made to you so you can check if the person sending you Bitcoins actually owned them.

It is important to note that an SPV is not as secured as a fully validating node because it usually contains the block headers and not the entire blocks. As a result, SPVs cannot validate transactions since they don't have them for a block and also because they do not have all the unspent transaction outputs (UTXOs) except for their own.

Network Discovery for a New Node

Now think about, when a new node wants to join the network, how would it contact the network? It is not an intranet with a 192.168.1.X network where you can broadcast to the IP 192.168.1.255 so that whichever computer is a part of the 192.168.1.X network gets the broadcast message. The network switches are designed to allow such broadcast packets. However, remember that we are talking about the Internet, which Bitcoin is sitting on. If you are running a node in London, there is a possibility that there are other nodes in London, Russia, Ireland, the United States, and India and all of them are connected through the Internet with some public facing IP address.

The question here is that when a fresh node joins the network, how does it figure out the peer nodes? There is no central server somewhere to respond to their request the way a typical Internet-based web application works. Blockchain is decentralized, remember? When started for the first time, a Bitcoin Core or BitcoinJ program does not have the IP address of any full node. So, they are equipped with several methods to find the peers. One of them is DNS seeds. Several DNS seeds are hardcoded in them. Also, several host names are maintained in the DNS system that resolve to a list of IP addresses that are running the Bitcoin nodes. DNS seeds are maintained by Bitcoin community members. Some community members provide static DNS seeds by manually entering the IP addresses and port numbers. Also, some community members provide dynamic DNS seed servers that can automatically get the IP addresses of active Bitcoin nodes that are running on default Bitcoin ports (8333 for *mainnet* and 18333 for *testnet*). If you perform NSLOOKUPs on the DNS seeds, you will get a bunch of IP addresses running Bitcoin nodes.

The clients (Bitcoin Core or BitcoinJ) also maintain a hardcoded list of IP addresses that point to some (not one!) stable Bitcoin nodes. Such nodes can be called bootstrap nodes whose endpoints are already available with the source code itself. Every time one downloads the binaries, a fresh list of active nodes get downloaded along with the binaries. Once a Bitcoin node connection is established, it is very easy to pull the list of other Bitcoin nodes active at that point in time. A pictorial representation of how a new node becomes a part of the network can be found in the following figures.

Step-1:

Imagine that there were six nodes active at some point in time in the Bitcoin network. Refer to Figure 3-10.

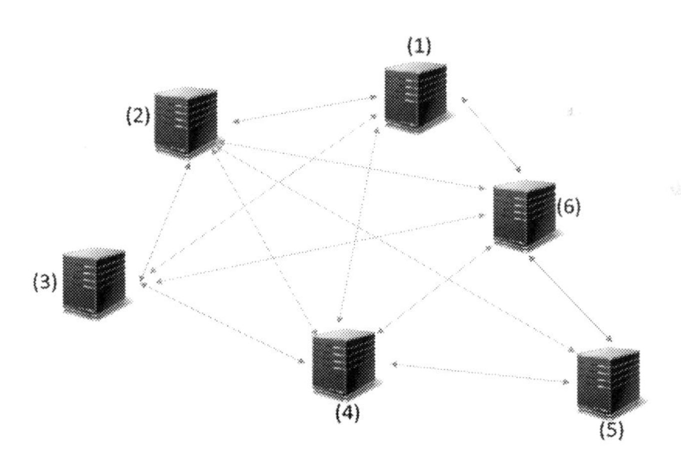

Figure 3-10. *Bitcoin network in general*

Step-2:

There is a new node, say, a seventh node that just showed up and is trying to join the existing Bitcoin network, but does not have any connection yet. Refer to Figure 3-11.

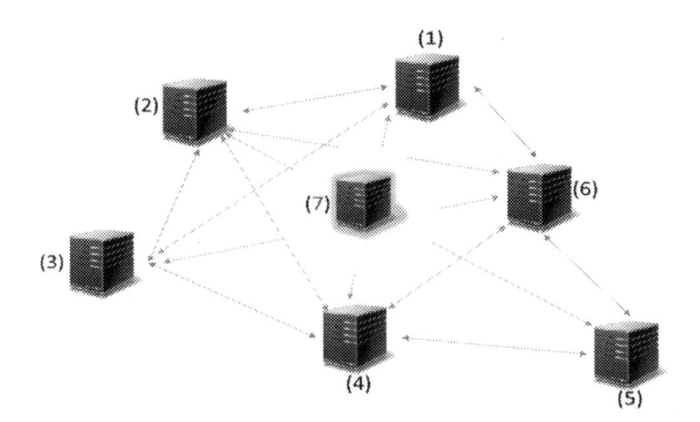

***Figure 3-11.** A new node trying to join the network*

Step-3:

The seventh node will try to reach out to as many nodes as it can either using DNS seeds or using the list of stable Bitcoin nodes in the list that it has–as shown in Figure 3-12.

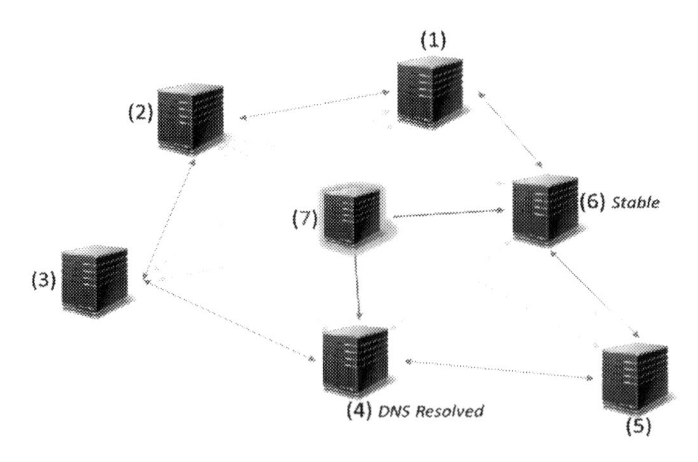

***Figure 3-12.** New Bitcoin node contacts some peers*

In the diagram, we have skipped the DNS resolution part. It is the same as when you browse any website with its name and post DNS resolution the IP address is retrieved, which is then used as the destination webserver's address to send TCP packets to. To connect to a new peer, the node establishes a TCP connection on port 8333 (port 8333 is well known for Bitcoins but could be different). Then the two nodes handshake with information such as version number, time, IP addresses, height of blockchain, etc. The actual Bitcoin code for "Version" message defined in *net.cpp* is as shown in the following:

```
PushMessage( "version", PROTOCOL_VERSION, nLocalServices,
nTime, addrYou, addrMe,
                nLocalHostNonce, FormatSubVersion(CLIENT_NAME,
                CLIENT_VERSION,
                std::vector<string>()), nBestHeight, true );
```

Through this Version message, the compatibility between the two nodes is checked as the first step toward further communication.

Step-4:

In the fourth step, the requested nodes will respond with the list of IP addresses and corresponding port numbers of the other active Bitcoin nodes that they are aware of. Please note that it is possible for some of active nodes to not be aware of each and every Bitcoin node in the network at any time. The port number is important because once the TCP packets reach the destination node, it is the port number that is used by the operating system to direct the message to the correct application/process running on the system. Please refer to Figure 3-13.

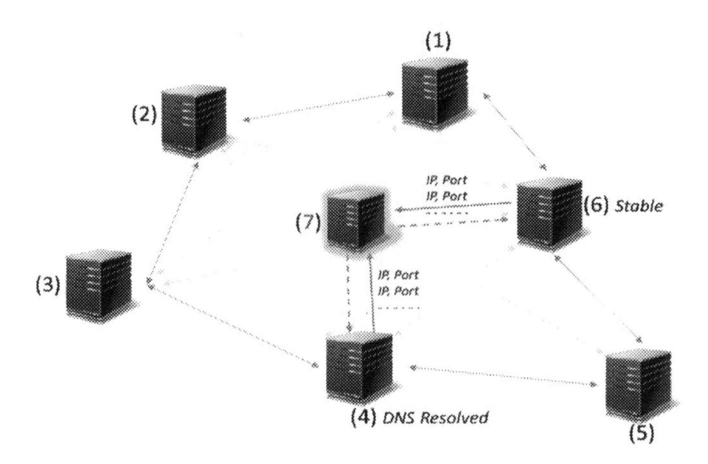

Figure 3-13. *Peer Bitcoin nodes respond to the network request by a new node*

Note that, only one peer may be enough to bootstrap the connection of a node to the Bitcoin network; the node must continue to discover and connect to new peers. This is because nodes come and go at will and no connection is reliable.

Step-5:

In the fifth step, the new seventh node establishes connection with all the reachable Bitcoin nodes, depending on the list it received from the nodes contacted in the previous step. Figure 3-14 represents this.

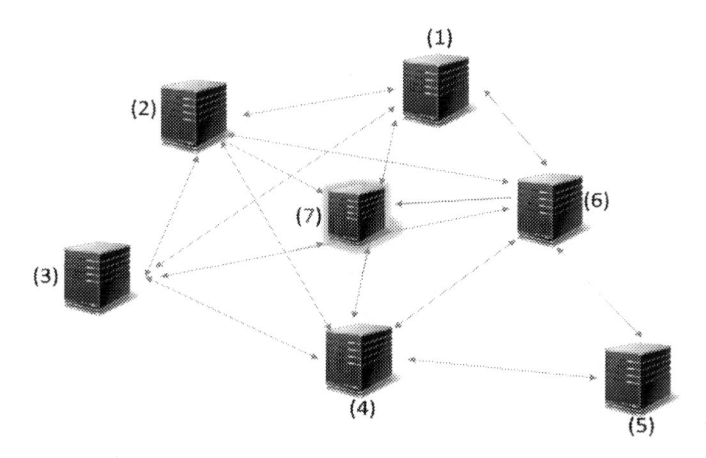

Figure 3-14. *A new node becomes a part of the Bitcoin network*

Bitcoin Transactions

Bitcoin transactions are the fundamental building blocks of the Bitcoin system. There are basically two broader categories of Bitcoin transactions:

- **Coinbase transaction**: Every block in Bitcoin blockchain contains one coinbase transaction included by the miners themselves to be able to mine new coins. They do not have control of how many coins they can mine in every block because it is controlled by the network itself. It started with 50 BTC in the beginning and keeps halving till it reaches 21 Million Bitcoins in total.
- **Regular transactions**: The regular transactions are very similar to currency exchanges in general, where one is trying to transact some amount of money that they own with another. Typically, in Bitcoin, everything is present as transactions. To spend some amount, one has to consume previous transaction(s) where they

received that amount—these are regular transactions in Bitcoin. Our main focus in this chapter will be on these regular transactions.

Each owner of a Bitcoin can transfer the coin to someone else by digitally signing a hash of the previous transaction where they had received the Bitcoin along with the public key of the recipient. The payee or the recipient already has the public key of the payer so they can verify the transaction. The following figure (Figure 3-15) is from the white paper of Satoshi Nakamoto that pictorially demonstrates how it works.

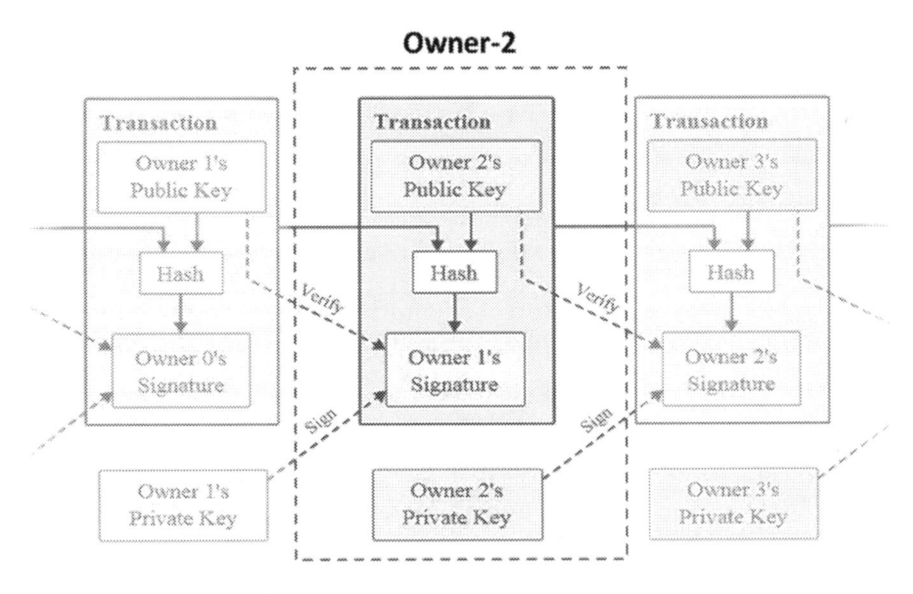

Figure 3-15. *Bitcoin transaction*

Notice only the highlighted Owner-2 section in the diagram. Since Owner-1 is initiating this transaction, he is using his private key for signing the hash of two items: one is the the previous transaction where he himself received the amount and the second is Owner-2's public key. This signature can be easily verified using the public key of Owner-1 to ensure that it is a legitimate transaction. Similarly, when Owner-2 will initiate a transfer to Owner-3, he will use his private key to sign the hash of the previous transaction (the one he received from Owner-1) along with the public key of Owner-3. Such a transaction can be, and will be, verified by anyone who is a part of the network. Obviously because every transaction is broadcast, most of the nodes will have the entire history of transactions to be able to prevent double-spend attempts.

There is no principle of closing balance in a Bitcoin network, and the total amount one holds is the summation of all incoming transactions to the public addresses you own. You can create as many public addresses as you want. If you have ten public addresses, then whatever transactions were made to that public address, you can spend those transactions (unspent transactions or UTXOs) using your private key. If you have to spend, say, five Bitcoins, you have a coupe of choices:

- Use one of the previous transactions where you received five or more Bitcoins. Transfer five Bitcoins to the recepient, some amount as transaction fee and the remainder to yourself. Refer to Figure 3-16.

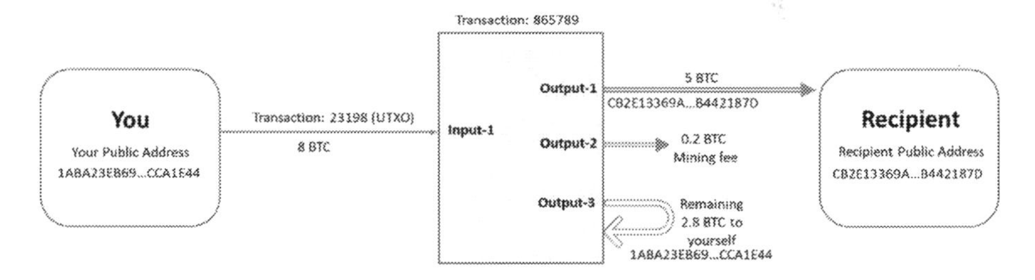

Figure 3-16. *Bitcoin transaction with one transaction input*

- Use multiple previous transactions that you had received that would sum up to more than five Bitcoins. Transfer five Bitcoins to the recepient, some amount as transaction fee and the remainder to yourself. Refer to Figure 3-17.

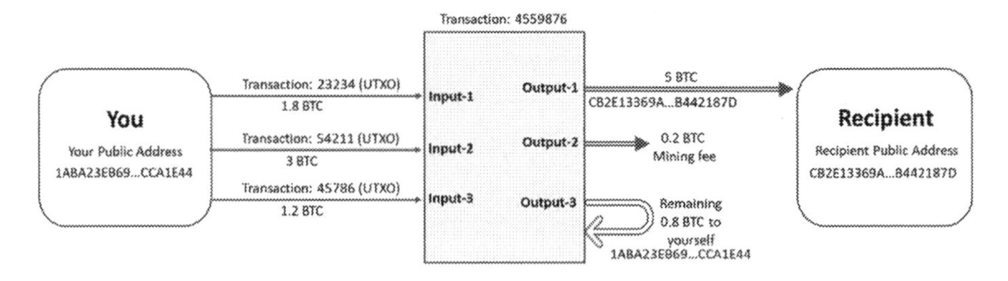

Figure 3-17. *Bitcoin transaction with multiple transactions input*

As you can see, every transaction takes as input the previous transaction(s). There is no account maintained that says you have eight BTC, and you can spend anything below this amount; if you spend five BTC, the remaining balance would be three BTC. In Bitcoin, everything is a transaction where there are inputs and outputs. If the outputs are not spent yet, they are the UTXOs.

We are aware that every transaction in the network is broadcast to the entire network. Whether someone is maintaining a node or not, they can still make a transaction and that transaction is published to all the accessible Bitcoin nodes. The receiver Bitcoin nodes then further broadcast the transactions to other nodes and the entire network is usually flodded with transactions. This is sometimes referred to as the gossip protocol and plays an important role in preventing double-spend attacks. Recollect from Chapter 2 that the only way to prevent double-spend is to be aware of all transactions.

Each node maintains a set of all the transactions that they hear about and broadcasts only the new ones, which were not a part of the list already. Nodes maintain the transactions in the list till the time the transaction gets into a block and is a part of the blockchain. This is because there is a chance that even if a block has all valid transactions and is proposed as a valid block, it can still get orphaned by not being a part of the longest chain. Once it is confirmed that the block is now a part of the longest chain, the transactions that are there in that block are taken off from the list of transactions. Each full node in the Bitcoin network must maintain the entire list of unspent transactions (UTXOs) even though they are in the millions. If a transaction is in the list of UTXOs, then it may not be a double-spend attempt. Upon confirming a transaction is not a double-spend attack and also validating the transactions from other perspectives, a node broadcasts such transactions. If you are wondering how fast it would be to search millions of UTXOs to check for double-spend, you are on track. Since the transaction outputs are ordered by their hashes, searching for an item in an ordered hash list is quite fast.

Let us now think and dig deeper into a double-spend scenario. It is very possible that Alice (A) tries to pay Bob (B) and Charlie (C) the same transaction (input to a transaction is a previous transaction and there is no concept of closing balance). Such a scenario would appear as shown in Figure 3-18.

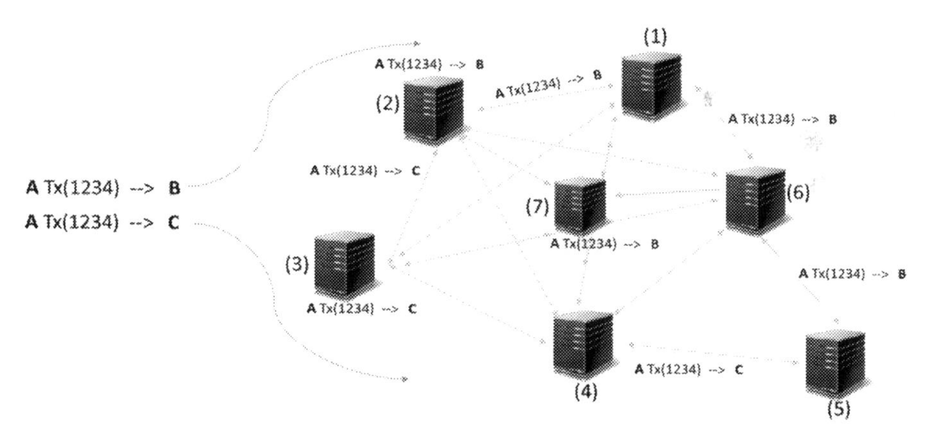

Figure 3-18. *A double-spend transaction scenario in Bitcoin network*

Notice in the figure the following scenarios:

- **A** is trying to spend the same transaction to **B** and **C**.
- Node-2 received the transaction **A** Tx(1234) --> **B** and Node-3 received the transaction **A** Tx(1234) --> **C**.

- For Node-2 and Node-3, their respective transactions received were legitimate transactions.
- When Node-3 tries to broadcast the transaction **A** Tx(1234) --> **C** to Node-2 (every node broadcasts new transactions), Node-2 would refuse this transaction because it already has a transaction **A** Tx(1234) --> **B** with the same input transaction Tx(1234).
- Similar things happen with other nodes as well, and they may have either the transaction "**A** Tx(1234) --> **B**" or "**A** Tx(1234) --> **C**", whichever reached them faster, but not both.
- During mining, whichever node gets to propose the block will include the transaction it has. This transaction would be a part of the blockchain and the rest of the nodes that are holding the other transaction would simply drop the transaction with Tx(1234) because it will no longer be a UTXO.

Consensus and Block Mining

In the previous section we looked at granular transactions. We will learn how the transactions are bundled together to form a block and a consensus is achieved among nodes, so the entire network accepts that block to extend the blockchain. Note that "block mining" refers to successfully creating a new block in the blockchain. In Bitcoin, it is the distributed PoW consensus algorithm that helps mine new blocks by maintaining decentralization. Achieving distributed consensus in such a network is very difficult. Though it has been around for decades for distributed systems such as Facebook, Google, Amazon, and many more because they have millions of servers that require consistency in the data they store, the term *consensus* is very much popularized because of Bitcoins. We will get into the nuts and bolts of consensus and mining in this chapter.

First, just keep in mind that everything in a Bitcoin network is represented as transactions. If you want to make a transaction, you have to consume one or more previous transactions as input and make another transaction. We already know that one has to sign a transaction using their private key to ensure that the right person is making the transaction. Despite such cryptographic security, can't that person sign a transaction that they already spent? Example: Alice received ten Bitcoins in a transaction with transaction number 1234. She can very well spend the same transaction 1234 and give away those ten Bitcoins to Bob and Charlie. Since she will sign with her private key, which means it is an authentic transaction, what do you think can prevent her from double-spending? Please note that there is no way in Bitcoin that you can prevent her from attempting to make a double-spend, but the system is designed so that such an attempt will not be successful. The only way to prevent such attempts is to be aware of all the transactions that are taing place. This is why all transactions in Bitcoin are broadcast to the entire network. Once a transaction is spent, it is no longer a part of the UTXOs and a new transaction number is generated to be a part of the UTXO, which only the recipient can spend. This is the way nodes can validate transactions.

Also, the only way you can prevent a double-spend attack is by knowing all the transactions. When you are aware of all transactions, you would know the spend and the UTXOs. When a new block is proposed by a miner, it is required that all the transactions in the block are valid. Does it mean that the node proposing a block cannot include a invalid transaction? The answer is "Yes." They can certainly inject a fraudulent transaction but the rest of the nodes will reject it. The PoW that the node would have done (we will get into the details shortly) by spending computer resources and electricity would be in vain! So, a node would never want to propose an invalid block, thanks to the PoW consensus.

Despite not having a notion of global time, observe that the transactions are clubbed together to form a block that becomes a part of blockchain, more blocks get added to the chain one by one, and there is an order! Note carefully that the order in which the transactions took place is preserved by the blockchain. This way, consensus happens at block level, which propagates all the way to granular transactions.

Based on what we have understood so far, now we know that every node in the Bitcoin network has its own copy of blockchain, and there is no "global blockchain" as such; it is a decentralized network after all. Each node in all those copies of blockchains comprises many transactions. It is also true that every node maintains a list of UTXOs and when given a chance (A node is randomly selected–we will see how) to propose a block, they include as many transactions as possible up to the block limit of 1MB or 2MB. If the block successfully makes it to the blockchain, they are removed from the list of UTXOs. Note here that every node may have different outstanding transactions lists because there is a possibility that some transactions are not heard by some of the nodes.

It's time now to see how the PoW algorithm really works. We learned about the difficulty target field in the header of each block. Every mining node tries to solve the cryptographic puzzle with an expectation to get lucky and propose a block. The reason they are so desperate in proposing a block is because they get great benefits when their proposed block becomes a part of the blockchain. Every transaction that an individual makes, they can set aside some transaction fee for the miners. We know that all nodes maintain the list of transactions that are not yet a part of the blockchain and when they get a chance to propose a block, they take as many transactions as they can and form a block. It is obvious that they will take all those transactions that would give them the highest profit and leave the ones with minimum or no transaction fees. It may take some time for the transactions with low transaction fee to get into the blocks, and chances are less for the ones with no transaction fee at all. Apart from the transaction fee, the nodes that propose a new block are rewarded with new Bitcoins. With every successful block, new Bitcoins get generated and the miner who proposed the block gets all of those; this is the only way new Bitcoins get created in the Bitcoin system. This is called "block reward." Technically, the node that proposes a block includes a special transaction called "coin creation" in the proposed block where the recipient address is the one that the miner owns. When the Bitcoin was first launched, the block reward was 50 Bitcoins (BTC). By design, there can only be 21,000,000 BTCs in total, so the block reward gets halved every 210,000 blocks. It started at 50, then it became 25, then 12.5, and it goes on this way till at some point in time (when it reaches 21,000,000 BTCs) it trends to zero. Following is the code snippet from Bitcoin Core (*main.cpp*) that shows this halving process:

```
int64_t GetBlockValue(int nHeight, int64_t nFees)
{
    int64_t nSubsidy = 50 * COIN;
    int halvings = nHeight / Params().SubsidyHalvingInterval();

    // Force block reward to zero when right shift is undefined.
    if (halvings >= 64)
        return nFees;

    // Subsidy is cut in half every 210,000 blocks which will
    occur approximately every 4 years.
    nSubsidy >>= halvings;

    return nSubsidy + nFees;
}
```

Notice in the previous code snippet how the block reward gets halved. The following explanation gives a better picture of this design:

```
//Block reward reduced by half, remains at 50%
BlockReward = BlockReward >> 1;

//Block reward further reduced by half, remains at 25%
BlockReward = BlockReward-(BlockReward>>2);

//Block reward further reduced by half, remains at 12.5%
BlockReward = BlockReward - (BlockReward>>3);
```

Even though the rewards look lucrative, it is not so easy to get lucky and be the node that gets to propose a block. If you are not lucky enough, all the work you did would be in vain; that's a disadvantage! So, what is it that the nodes do as a PoW? Let's get back to the *difficulty* puzzle now. Every mining node at all times is working to propose a block, but only one succeeds at a given point in time. Assume that a block is proposed already, and now all mining nodes are working to propose a new block. Let us go through the process step by step and understand the whole flow:

Step-1:

The miners use a software to keep track of the transactions, eliminate the ones that already made it to a successful block in blockchain, reject the fraudulent transactions, and solve the cryptographic puzzle to propose a new block and relay that to the entire network. The best software to mine is the official Bitcoin Core but there have been many other variants that people have come up with. If you log into this link (`https://bitcoin.org/en/download`) you will find that the official Bitcoin Core is supported in Windows, Linux, Mac OS, Ubuntu, and ARM Linux. So, when we say that the mining node selects all the transactions (maybe the ones that give the miner the highest profit) till the block limit of 1MB (2MB for Bitcoin Cash), they also hash those transactions and generate the Merkle root that would become a part of this new block's header. This Merkle root represents all the transactions.

Step-2:

They prepare the block header. Apart from the nonce, the rest is all available at this step. It is the work of the mining node to find the nonce by hashing the block header twice and comparing it against the difficulty target to see if it is less than that. They keep changing the nonce till it satisfies this condition, and there is no shortcut to find a nonce quickly; one must try out every possible option. We already looked at how to compute the difficult target using the four bytes of data present in the header itself, and we learned how it changes every two weeks. See the following for how this process looks:

```
H [ H (Version | Previous Block Hash | Merkle Root | Time Stamp
| Difficulty Target |  Nonce) ]

< [ Difficulty Target ]
```

Step-3:

The miner keeps on changing the nonce field in step-2, by incrementing it by "1" till it satisfies the condition–it is a recursive phenomenon. The difficulty target for every node is the same and all of them are trying to solve the same problem, but remember that they may have different variants of transaction pools and hence the Merkle root for them would be different. Since every node is trying to extend the longest and main blockchain, so the previous block hash would be the same.

So, ultimately the Sha256 hash twice for the block header should be less than the target to be able to propose the block to the entire network. See the following example for a better understanding:

```
Target  : 0000000000000074cd0000000000000000000000000000000000
          0000000000000

Hash    : 0000000000000074cc4471deff052ced7f07347e4eda86c84
          5a2fcf0553ed7f0
```

Notice that the hash value and the target value have the same number of leading zeros (i.e., 14) and "74cc" is less than "74cd," so it satisfies the condition and this block can now be proposed. In many places, you would find that this explanation is simplified with ballpark values of both the *target* and the *hash,* and counting only the leading zeros. If the hash has more leading zeros than the target, then it satisfies the condition. Remember again that the more zeros in the target, the more difficult it gets in finding the hash that can satisfy the condition.

Let us connect this learning so far with the real Bitcoin implementation. We know that block creation time is set to ten minutes–it is coded up in Bitcoin binaries for 2,016 blocks in two weeks, as we discussed already, and does not change till a hard fork happens. You can browse blocks proposed and see the hashes that satisfied the difficulty target at the website `https://blockchain.info` and see for yourself that the hashes for different blocks would have different leading zeros, just to set the block creation time to ten minutes on average. In the initial days, the number of leading zeros was around nine or ten, and today it has increased to around 18 to 20 zeros. It may increase even further as and when more powerful computing nodes capable of more hash rates join the network.

Step-4:

Once a miner finds the valid block, they immediately publish the block to the entire network. Every node that receives this block individually checks again if the miner who proposed the block actually solved the mining puzzle. For these nodes to validate this, it is just one step, as shown below:

```
H [H (Version| PreviousBlock Hash | Merkle Root | Time Stamp |
Difficulty Target | Found Nonce)]

< [ Difficulty Target ]
```

Notice that they just use the block header that includes the nonce found by the proposing miner to see if the hash is less than the target and is valid. If it was a valid nonce and the condition satisfied, then they check for individual transactions proposed in the block with its Merkle root in the block header. If all transactions are valid, then they add this block to the local copy of their blockchain. This new block has the coinbase transaction that generates new coins (it started with 50 BTC, then 25, then 12.5, and keeps halving as discussed) as an award for the miner who proposed the valid block.

Note here that block mining is not an easy job, thanks to the PoW mining algorithm. For a node to propose an invalid block, it has to burn a lot of electricity and CPU cycles to find the nonce and propose the block that would ultimately get rejected by nodes in the network. Had it been an easy task, many nodes would just keep trying for it and flood the network with bad blocks. You must understand and appreciate by now how Bitcoin prevents such situations in a game theoretic way! It is always profitable for the miners to play by the rules and they do not gain any extra benefits by not following the rules.

In the previous steps, we learned the PoW mining procedure that is implemented in Bitcoins. One of the best things in this design is the random selection of a mining node that gets to propose a block. No one knows who would get lucky to find the right nonce and propose a block, it is purely a random phenomenon. We already know that generating a true random number is quite difficult and this is something that is the most vulnerable surface of attack for most cryptographic implementations. With such a design as Bitcoin's, random selection of a node to propose a block is truly random.

The next best thing in Bitcoin mining is the block reward. It is something that a miner who successfully proposes a new block gets, using the coinbase transaction in the same block. The miners also get the transaction fees associated with all the transactions they have included in the block. So, mining reward for a block is a combination of block reward and transaction fee as shown below:

```
Mining Reward = Block Reward + Total transaction fees of all transactions in the Block
```

We know that mining is the only way new Bitcoins are created in the Bitcoin system, but is that the purpose of mining? No! The purpose of mining is to mine new blocks, and generation of new Bitcoins and also the transaction fee is to incentivize the miners so that more and more miners are interested in mining. Obviously though, why would you mine if you are not making good money? This is again game theory. A proper incentivization mechanism is the key to make a sytem decentralized and self-sustainable. Notice that the Bitcoin system does not have a way to penalize nodes that do not play honestly, it only rewards honest behavior. Actors in the Bitcoin blockchain network such as individuals who just use Bitcoin or the miners are all identified using their public keys. It is possible for them to generate as many key pairs as possible and this makes it a psydonemous system. A node cannot be uniquely identified with its public key that it has used in the coinbase transaction, as in the very next moment it can create a new key pair and expose itself with a new network address. So, proper incentivization stands to be the best way to ensure the actors in the system play honestly–again the beauty of game theory!

Here is a question for you now. After a block was broadcast, let's say a node verified it, found the nonce and transactions and everything else to be valid, and included it in its local copy of blockchain. Does this mean that the transactions that were there in the block are all settled and confirmed now? Well, not really! There is a chance that two blocks came in at the same time and while one node started extending one of them, there is a chance that a majority of the nodes are extending on the other block. Eventually, the longest blockchain becomes the original chain. This is a scenario when a block that is absolutely valid, with all legitimate transactions and a proper nonce value that satisfied the mining puzzle, can still get abandoned by the Bitcoin network. Such blocks that do not become a part of the final blockchain are called orphaned blocks. Now, this explanation indicates that there is a certain possibility of one or more blocks getting orphaned out at any time. So, the best strategy would be to wait untill many blocks are added to the chain. In other words, when a transaction receives multiple confirmations, it is safe to think that it is a part of the final consensus chain and will not get orphaned out. As any number of blocks get added after a certain block, that many number of confirmations are received by the transactions in that block. Though there are no rules as such that define how many confirmations one should get before accepting a transaction, six confirmations has been the best practice. Even with four confirmations, it is quite safe to assume a transaction has been confirmed, but six is the best practice because with more confirmations, the chances of a block getting orphaned out decreases exponentially.

Block Propagation

Bitcoin uses PoW mining to randomly select a node that can propose a valid block. Once the miner finds a valid block, they broadcast that block to the entire network. Block propagation in the network happens the same way as transactions. Every node that receives the new block further broadcasts it so that eventually the block reaches every node in the network. Please note that a node does not broadcast a block unless it is a part of the longest chain from its perspective. Once the nodes receive the new block that is proposed, they not only verify the header and check the hash value in acceptable range, but also validate each and every transaction that was included in that block. It is clear by now that for a node in the Bitcoin network, validating a block is a little more complex compared with validating transactions. Similar to transactions, there is a possibility that two valid blocks are proposed at the same time. In such a scenario, the node will keep both the blocks and start building on the one that comes from the longest chain.

We must understand that there is always a latency involved for a block to propagate through the entire network and reach every node. The relation between the block size and the time taken is linearly proportional, in the sense that for each kB added to the block size, the latency increases linearly. It is obvious that such network latency would impact the rate of growth of the blockchain. A measurement study that was conducted by Decker and Wattenhofer addresses this situation. Refer to Figure 3-19, which shows the relation between block size and the time it took to reach 25% (Line-1), 50% (Line-2), and 75% (Line-3) of monitored nodes.

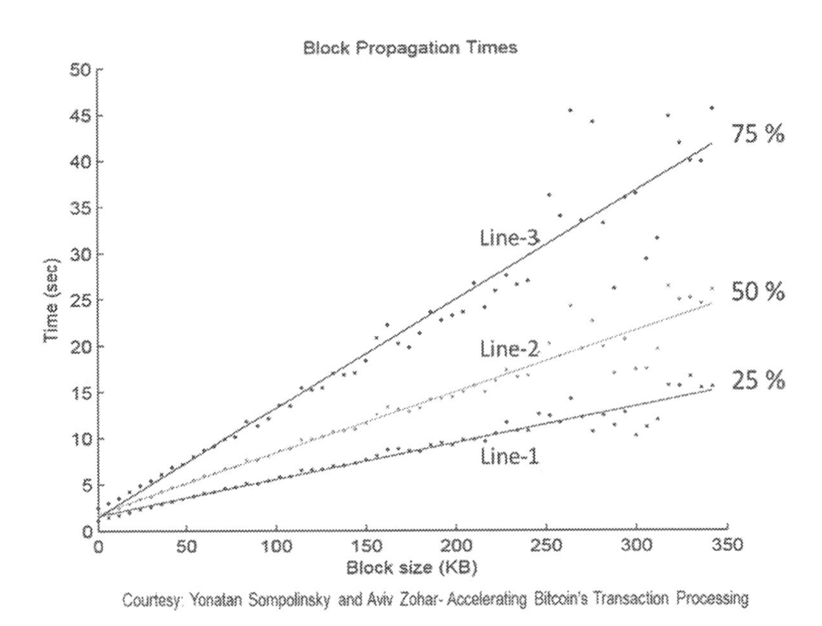

***Figure 3-19.** Block propagation time with respect to block size*

The network bandwidth is the primary reason for such network latencies and it is never consistent in all areas of the globe. On top of this, we know that the broadcast packets of blocks go through many hops to finally reach all nodes. A typical Bitcoin block is of 1MB and the new variant of Bitcoin with a hard fork that has come up (Bitcoin Cash) is of 2MB block size; you can imagine the inherent limitations due to latency.

As per the network research, there are more than a million Bitcoin nodes that are connected to the Bitcoin network in a month and there are thousands of full nodes that are almost always connected to the network permanently.

Putting It all Together

At a high level, if we just put down the events in the order they take place, then here is how it may look:

- All new transactions are broadcast to all nodes.
- Each node that hears the new transactions collects them into a block.
- Each mining node works on finding a difficult PoW for its block to be able to propose it to the network.
- When a node gets lucky and finds a correct nonce to the PoW puzzle, it broadcasts the block to all nodes.
- Nodes accept the proposed block only if the nonce and all transactions in it are valid and not already spent.
- Bitcoin network nodes express their acceptance of the block by working on creating the next block in the chain, using the hash of the accepted block as the previous hash for the new block they would be mining.

Bitcoin Scripts

We learned about Bitcoin transactions in previous sections at a high level. In this section we will delve deep into the actual programming constructs that make the transactions happen. Every Bitcoin transactions' input and output are embedded with scripts. Bitcoin scripts are stack based, which we will see shortly, and are evaluated from left to right. Remember that Bitcoin scripts are not Turing-complete, so you cannot really do anything and everything that is possible through other Turing-complete languages such as C, C++, or Java, etc. There are no concepts of loops in Bitcoin script, hence the execution time for scripts is not variable and is proportional to the number of instructions. This means that the scripts execute in a limited amount of time and there is no possibility for them to get stuck inside a loop. Also, most importantly, the scripts definitely terminate. Now that we know a little about the scripts, where do they run? Whenever transactions are made, whether programmatically, or through a wallet software or any other program, the scripts are injected inside the transactions and it is the work of the miners to run those scripts while mining. The purpose of Bitcoin scripts is to enable the network nodes to ensure the available funds are claimed and spent only by the entitled parties that really own them.

Bitcoin Transactions Revisited

A transaction in the Bitcoin network is a tranfer of value, which is a broadcast to the entire network and ends up in some block in the blockchain. Typically, it appears that Bitcoins get transferred from one account or wallet to another, but in reality, it is from one transaction to another. Well, before getting into further details, keep in mind that the Bitcoin addresses are basically the double-hashed output of the public key of the participants. The public key is first hashed using SHA256 and then by RIPEMD160 hashing algorithms, respectively, to generate 160-bit Bitcoin addresses. We have already covered these hashing techniques in the previous chapter. Let us zoom in a bit more into the transactions now. Take a look at the following transaction tree (Figure 3-20), the way it happens in Bitcoin.

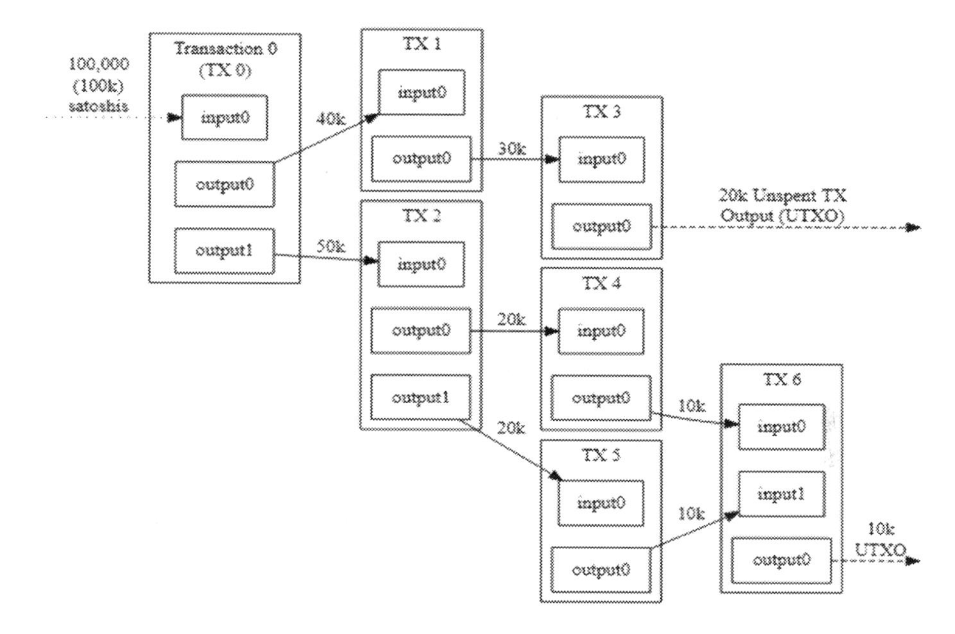

Figure 3-20. *A typical Bitcoin transaction structure*

Observe that the output of previous transactions become input to new transactions and this process continues forever. In the previous figure, if it was you who got the 100K from some previous output, it became the spendable input to a new transaction. Notice that in Tx 0, you spent 40K and 50K and paid up those amounts, and the remaining amount (10K) became the fee to the miner. By default, the remaining amount is paid to the miner, so you need to be careful not to ignore such situations, which are always the case. In this same situation, out of the remaining 10K amount, you could transfer say 9K to your own address and leave aside 1K for the mining fee. When an amount is not spent, in the sense that a transaction is not used as an input to a new transaction, it remains a UTXO, which can be spent later. Obviously, the ones previous to it are already spent. So, all the UTXOs combined for all the accounts (public keys) that you hold are your wallet balance.

Pause for a moment here, and think about how it must have been programmed. Remember that both the inputs and outputs of transactions are equipped with relevant scripts to make it possible. It is only through the scripts that it can be ensured that you are the authorized user to make a transaction and you have the necessary amount that you have received from a previous transaction. This means that both the inputs and outputs are equally important. Here is how the transaction's contents look:

```
Transaction Output = Amount of Bitcoins to transfer + Output
Script
Transaction Input = Reference to previous transaction output +
Input Script
```

Whether to look into the output script first or the input script first is actually an egg-chicken problem. But we will see the output script first because it is the one that is being consumed by the input script of the next transaction. Let us repeat and get this right, that while making a transaction, the output script of the current transaction is there just to enable the future transaction that can consume this as input, but for this current transaction, it is the previous transaction's output script that lets you spend it. This is why the output scripts have the public key of the recipient and the value (amount of Bitcoins) being transferred. When the output scripts are being used as inputs, their primary purpose is to check the signature against the public key. The output scripts are also called *ScriptPubKey*. Unless this output is spent, it remains a UTXO waiting to be spent.

The input script in the transaction input data structure has the mechanism of how to consume the previous transaction that you are trying to spend. So, it has to have the reference to that previous transaction. The hash of the previous transaction and the index number {hash, index} pinpoints the exact place where you had received the amount that you are now spending. The purpose of the "index" is to identify the intended output in the previous transaction. If you were the recipient of the previous transaction, you have to provide your signature to claim that you are the rightful owner of the public key to which the transaction was made. This will let you spend that transaction. Also, you have to provide your public key, which will hash to the one used as destination address in previous transaction. Input scripts are also known as *ScriptSigs.* The ultimate objective of the script is to push the signatures and keys onto the stack.

A typical Bitcoin transaction has the following fields (Table 3-5).

Table 3-5. *Bitcoin Transaction Fields*

Field	Size	Description
Version no	4 bytes	Currently 1. It tells Bitcoin peers and miners which set of rules to use to validate this transaction.
In-counter	1 - 9 bytes	Positive integer (VI = VarInt). It indicates total number of inputs.
list of inputs	Variabe length	It lists all the inputs for a transaction.
Out-counter	1 - 9 bytes	Positive integer (VI = VarInt). It indicates total number of outputs.
list of outputs	Variable length	It lists all the outputs for a transaction.
lock_time	4 bytes	Not being used currently. It indicates if the transaction should be included in the blockchain block immediately after it is validated by the miner or there should be some lock time before it gets included in the block.

Let us now take a look at a different representation of the same transaction structure that we discussed in the previous section. This is to see a more detailed view of the transaction structure and the various components of it. Now refer to Figure 3-21.

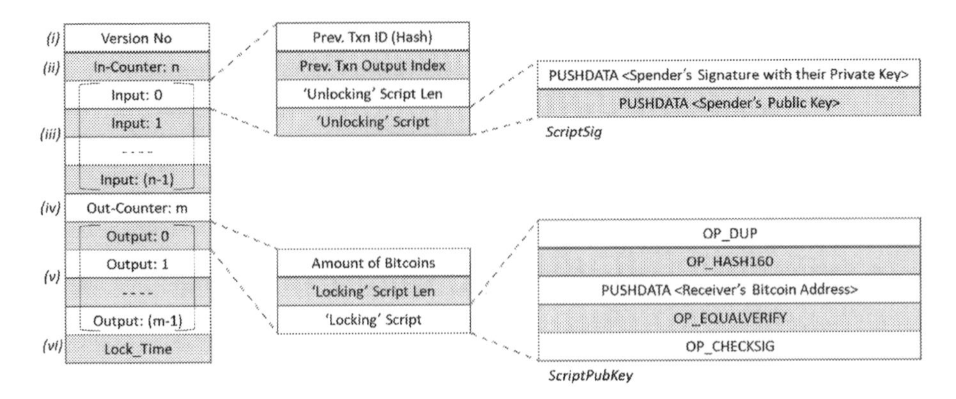

Figure 3-21. *Granular components of a Bitcoin transaction*

As you can see in the previous figure, the data items such as signatures or public keys are all embedded inside the scripts and are a part of the transaction. Just by looking at the granular components of Bitcoin transactions, many of your queries would be answered up front. The instructions in the script get pushed onto the stack and executed, which we wil explore in detail shortly.

When Bitcoin nodes receive the transaction broadcasts, they all validate those transactions individually, by combining the output script of the previous transaction with the input script of the current transaction following the steps mentioned as follows:

- Find the previous transaction whose output is being used as input to make this transaction. The "Prev. Txn ID (Hash)" field contains the hash of that previous transaction.
- In the previous transaction output, find the exact index where the amount was received. There could be multiple receivers in a transaction, so the index is used to identify the initiator of this current transaction whose address was used as recipient in the previous transaction.

- Consume the output script used in the previous transaction using the Unlocking Script called "*ScriptSig*." Notice in Figure 3-21 that there is a field before it that specifies the length of this Unlocking Script.
- Join this output script with the input script by just appending it to form the validation script and execute this validation script (remember this is a stack-based scripting language).
- The amount value is actually present in the Output Script, i.e., the "*ScriptPubKey*." It is the locking script that locks the transaction output with the spending conditions and ensures that only the rightful owner of the Bitcoin address to which this transaction has been made can later claim it. Observe that it also has the Locking Script Length field right before it. For the current transaction, this output script is only for information, and plays its role in the future when the owner tries to spend it.
- It is the validation script that decides if the current transaction input has the right to spend the previous UTXO by validating the signatures. If the validation script runs successfully, it is confirmed that the transaction is valid and the transaction went through.

Let us explore the previous explanation through a diagramatic representation to get a better understanding. Assume that Alice is paying Bob, say, five BTC. This means that Alice had received 5BTC in one of the previous transactions that was locked using *ScriptPubKey*. Alice can claim that she is the rightful owner of that transaction by unlocking it with *ScriptSig* and can spend it to Bob. Similarly, now if Bob tries to spend three

BTC to Charlie and two BTC to himself, then here is how it would look (Figure 3-22).

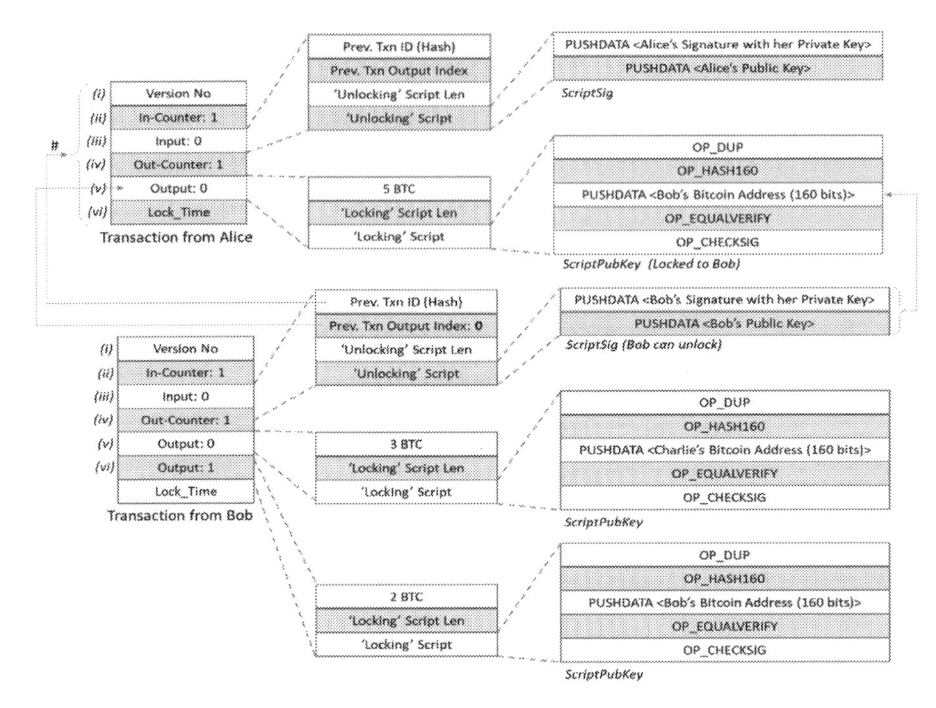

***Figure 3-22.** A practical example of Bitcoin scripts*

When the Bitcoin network, more precisely the miners, receive the transaction from Alice, they check and confirm that it is a valid transaction and approve it by including it in their blocks (well, the one who proposes a block does it). When that happens, the output of this transaction becomes a part of the UTXO in the name of Bob, who could later spend it. And this is what happens in our example–that Bob also spends it to Charlie. Bob did so by consuming the previous transaction, unlocking it with his signature and public key, to prove that he is the owner of the Bitcoin address that Alice had used. Observe that there are two outputs in Bob's transaction. Since he had received five BTC from Alice and is paying three BTC to

Charlie, he must transfer the remainder to himself so that it becomes two BTC of UTXO bound to Bob himself and he could spend it in future. In Bob's transaction, the three BTC to Charlie is locked using the locking script for only Charlie to spend later.

Are you now thinking about how the scripts are combined and executed together? Remember that the unlocking script of current transaction is run along with the locking script of the previous transaction. As discussed already, running the scripts is a miner's job and they do not happen at the wallet software. In the previous example, when Bob makes the transactions, miners execute the *ScriptSig* unlocking script from Bob's transaction, and then immediately execute the *ScriptPubKey* locking script from Alice's transaction in order. If the sequential execution in a stack-based fashion for the combined validation script runs successfully by returning TRUE, then Bob's transaction is excepted by all the nodes validating it. We will take a look at Bitcoin scripts and how a Bitcoin-scripting virtual machine executes the stack during the execution of the combined script commands in more detail in the following section. In this section, however, take a look at the following example that represents the transaction from a developer's standpoint:

Example code with just one input and one output

```
{
  "hash": "a320ba8bbe163f26cafb2092306c153f87c1c2609b25db0
  c13664ae1afca78ce",
  "ver": 1,
  "vin_sz": 1,
  "vout_sz": 1,
  "lock_time": 0,
  "size": 51,
```

```
  "in":[
      {
         "prev_out":{
            "hash":"83cd5e9b704c0a4cb6066e3a1642b483adc8f73a76
            791c82a73dfa381281d32f",
            "n":0
         },
         "scriptSig":"63883d3d2dea35029d17d25b8a926675def004
         5c397d3df55b0ae145ef80db7849599b930220ab13bd2dda2
         ca0a67e2c5cd28030bb9b7b3dcacf176652dac82fe9d5873
         f3409661281d32f6d35b46906cd562bf8b48f4f938c077bcb
         29d46b0560fa5c61813d3d2d"
      }
   ],
  "out":[
      {
         "value":"0.08",
         "scriptPubKey":"OP_DUP OP_HASH160 b3a2c0d84ec82cff
         932b5c3231567a0d48ab4c78
OP_EQUALVERIFY OP_CHECKSIG"
      }
   ]
}
```

Note that Bitcoin transactions are not encrypted, so it is possible to browse and view every transaction ever collected into blocks.

Scripts

A script is actually a list of instructions recorded with each transaction that describes how the next person can gain access to those Bitcoins received and spend them. Bitcoin uses stack-based, non–Turing-complete scripting language where the instructions get processed from left to right. Keep in mind that it is non-Turing-complete by design!

We looked at the input and output scripts in the previous section. We are now aware that the input script *ScriptSig* is the unlocking script and has two components, the public key and the signature. The public key is used because it hashes to the Bitcoin address that the transaction was spent to, in the previous transaction. The ECDSA digital signature's purpose is to prove the ownership of the public key, hence the Bitcoin address to be able to spend it further. Similarly, the output script *ScriptPubKey* in the previous transaction was to lock the transaction to the rightful owner of the Bitcoin address. These two scripts, *ScriptSig* of current transaction and *ScriptPubKey* of previous transaction, are combined and run. Take a look at its appearance after they are combined (Figure 3-23).

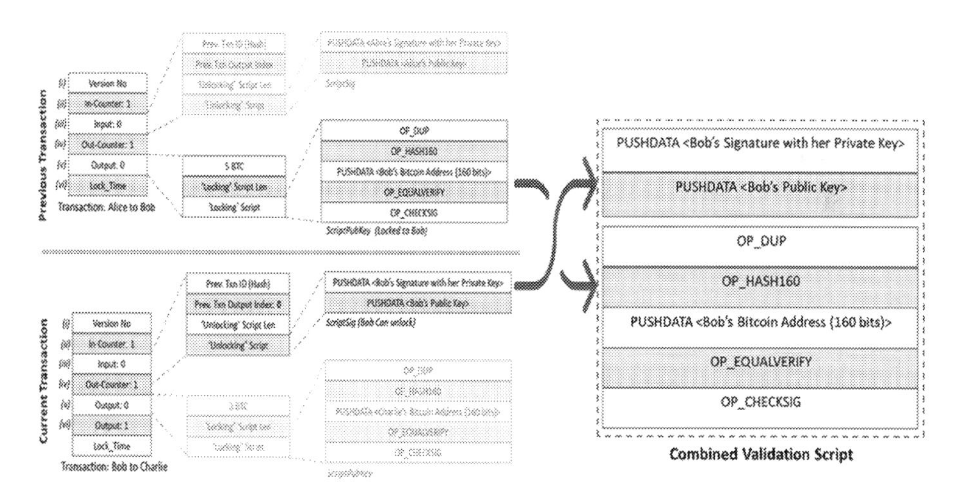

***Figure 3-23.** Formation of combined validation script*

As we learned already, it is important to note that the Bitcoin script either runs successfully or it fails. When the transactions are valid, it runs successfully without any errors. Bitcoin scripting language is a very simplified version of programming languages and is quite small, with just 256 instructions in total. Out of these 256, 15 are disabled and 75 are kept reserved maybe for later usage. These basic instructions comprise mathematical, logical (if/then), error reporting, and just return statements. Apart from these, there are some additional cryptographic instructions such as hashing, signature verification, etc. We will not get into all the available instruction sets, and focus only on the ones we will use in this chapter. Following are a few:

- **OP_DUP**: It just duplicates the top item in the stack.
- **OP_HASH160**: It hashes twice, first with SHA256 and then RIPEMD160.
- **OP_EQUALVERIFY**: It returns TRUE if the inputs are matched, and FALSE otherwise and marks the transaction invalid.
- **OP_CHECKSIG**: Checks if the input signature is a valid signature using the input Public Key itself for the hash of the current transaction.

To execute these instructions, we just have to push these instructions on to the stack and then execute. Apart from the memory that the stack takes, there is no extra memory required and this makes the Bitcoin scripts efficient. As you have seen, there are two kinds of instructions in the script, one is data instruction and the other is opcodes. The previous bullet list entries are all opcodes, and the combined validation script that we saw before has both of these kinds of instructions. Data instructions are just to push data onto the stack and not really to perform any function, and the sole purpose of opcodes is to execute some functions on the data in the stack and pop out as applicable. Let us discuss how Bob's transaction would get executed with such a stack-based implementation. Recollect the combined script where Bob is trying to spend a previously received transaction in the current transaction to Charlie (Figure 3-24).

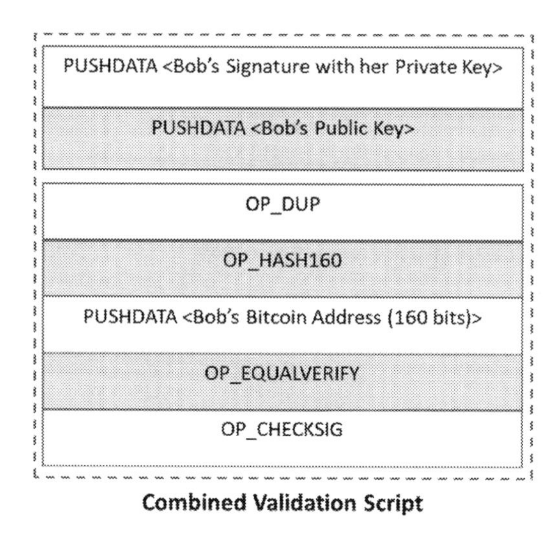

***Figure 3-24.** Combined script of ScriptPubKey and CheckSig*

The corresponding stack-based implemention would be as follows (Figure 3-25).

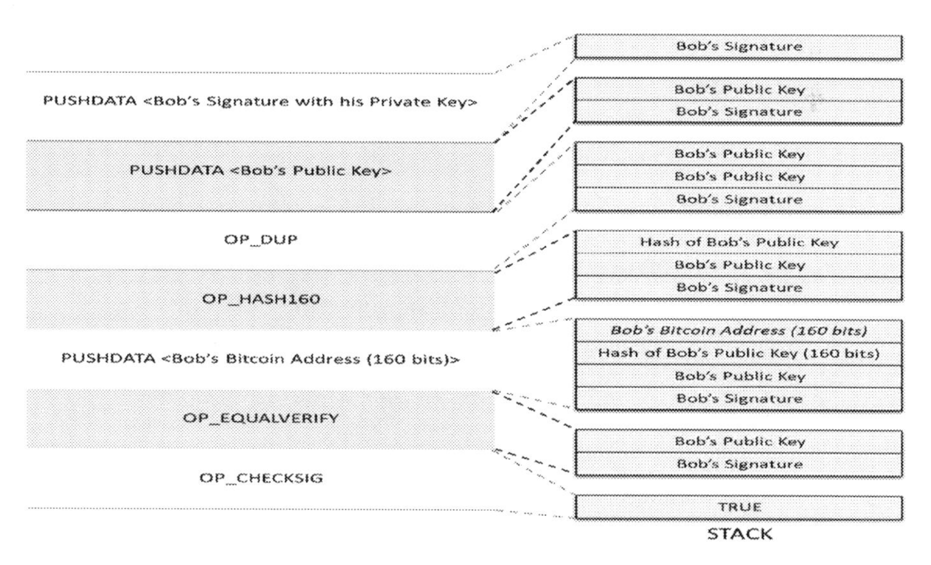

***Figure 3-25.** Example of stack-based implementation of Bitcoin script*

Though the previous stack-based implementation is self explanatory, we will quickly run through what happened here.

- First was Bob's signature–a data instruction and so was pushed onto the stack
- Then was his public key–again a data instruction and was pushed on to the stack
- Then it was OP_DUP, an opcode. It duplicates the first item in the stack, so the public key of Bob was duplicated and became the third item on the stack.
- Next was OP_HASH160, an Opcode, which hashed Bob's public key twice, once with SHA256 and then with RIPEMD160, and the final 160 bits output replaced Bob's public key and became the top of the stack.
- Then it was Bob's Bitcoin address (160 bits)–a data instruction, which was pushed to the stack.
- Next was an opcode, OP_EQUALVERIFY, which checks the top two items in the stack and if they match, it pops them both else an error is thrown and the script would terminate.
- Then was again an opcode OP_CHECKSIG, which checks the public key against the signature to validate the authenticity of the owner. This opcode is also capable of consuming both inputs and popping them off the stack.

You must be wondering what if someone tries to inject some fraudulent scripts or tries to misuse them. Please note that Bitcoin scripts are standardized and the miners are aware of them. Anything that does not follow the standard gets dropped by the miners, as they wouldn't waste their time executing such transactions.

Full Nodes vs. SPVs

We already got a heads-up on the full nodes and SPVs in this chapter. It is obvious that the notion of full node and lightweight node is implemented to ease out the usage of Bitcoins and make them more adaptable. In this section, we will zoom in to the technicalities for these variants and understand their purpose.

Full Nodes

The full nodes are the most integral components in a Bitcoin network. They are the ones that maintain and run Bitcoin from various places across the globe. As discussed already, download the entire blockchain with all transactions, starting all the way from the genesis block to the latest discovered block. The latest block defines the height of the blockchain.

The full nodes are extremely secure because they have the entire chain. For an adversary to be successful in cheating a node, an alternative blockchain needs to be presented, which is practically impossible. The true chain is the most cumulative PoW chain, and it gets computationally infeasible to propose a new fraudulent block. If all transactions are not valid in a block, PoW mining performed by the adversary will be in vain, because other miners will not mine on top of it. Such a block gets orphaned out soon enough. Full nodes build and maintain their own copy of blockchain locally. They do not rely on the network for transaction validation because they are self-sufficient. They are just interested in knowing the new blocks that get proposed by other nodes so that they can update their local copy after validating blocks. So, we learned that each full node must process all transactions; they must store the entire database, every transaction that is currently being brioadcast, every transaction that is ever spent, and the list of UTXOs; participate in maintaining the entire Bitcoin network; and they also have to serve the SPV clients.

Note that there are so many varities of Bitcoin software that the full nodes use that are quite different in software architecture and programmed in different language constructs. However, the most widely used one is the "Bitcoin Core" software; more than three fourths of the network uses it.

SPVs

Bitcoin design has this nice concept of Simple Payment Verification(SPV) nodes that can be used to verify transactions without running full nodes. The way SPVs work is that they download only the header of all the blocks during the initial syncing to the Bitcoin network. In Bitcoin, the block headers are of 80 bytes each, and downloading all the headers is not much and ranges to a few MBs in total.

The purpose of SPVs is to provide a mechanism to verify that a particular transaction was in a block in a blockchain without requiring the entire blockchain data. Every block header has the Merkle root, which is the block hash. We know that every transaction has a hash and that transaction hash can be linked to the block hash using the Merkle tree proof which we discussed in the previous chapter. All the transactions in a block form the Merkle leafs and the block hash forms the Merkle root. The beauty of the Merkle tree is that only a small part of the block is needed to prove that a transaction was actually a part of the block. So, to confirm a transaction an SPV does two things. First, it checks the Merkle tree proof for a transaction to accertain it is a part of the block and second, if that block is a part of the main chain or not; and there should be at least six more blocks created after it to confirm it is a part of the longest chain. Figure 3-26 depicts this process.

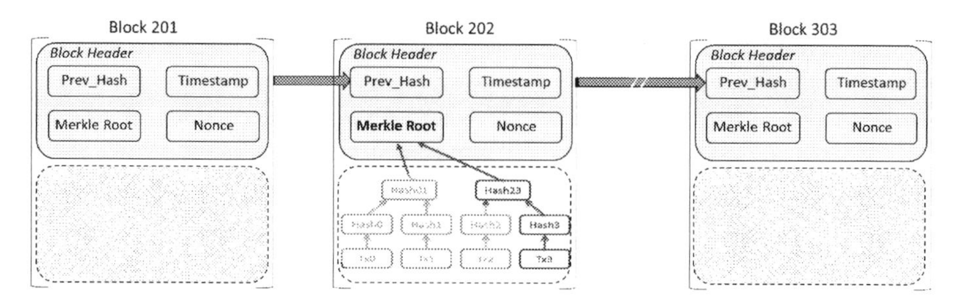

***Figure 3-26.** Merkle root in block header of SPV*

Let us dig deeper into the technicality of how SPVs really work in verifying transactions. At a high level, it takes the following steps:

- To the peers an SPV is connected to, it establishes Bloom filters with many of them and ideally not just to one peer, because there could be a chance for that peer to perform denial of service or cheat. The purpose of Bloom filters is to match only the transactions an SPV is interested in, and not the rest in a block without revealing which addresses or keys the SPV is interested in.
- Peers send back the relevant transactions in a *merkleblock* message that contains the Merkle root and the Merkle path to the transaction of interest as shown in the figure above. The *merkleblock* message sze is in a few kB and quite efficient.
- It is then easy for the SPVs to verify if a transaction truly belongs to a block in the blockchain.
- Once the transaction is verified, the next step is to check if that Block is actually a part of the true longest blockchain.

The following (Figure 3-27) represents this SPV communication steps with its peers.

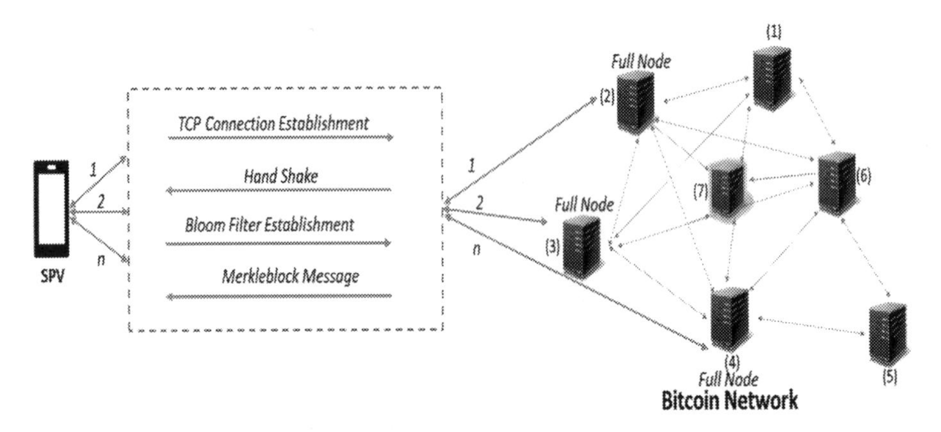

***Figure 3-27.** SPV communication mechanism with the Bitcoin network*

Bitcoin Wallets

Bitcoin wallets are very similar to the wallet you use in your daily life, in the sense you have access to it and you can spend when you want. Bitcoin wallets, however, are a digital phenomenon. Recollect the example we used in the previous section, where Alice paid some amount to Bob. How would she do it if Bob did not have an account? In the Bitcoin setting, the accounts or wallets are represented by the Bitcoin address. Bob must first generate a key pair (private/public keys). Bitcoin uses the ECDSA algorithm with *secp256k1* curve (don't worry, it is just the curve type–a standard recommendation). First a random bit string is generated to serve as private key, which is then deterministically transformed to public key. As we learned before in Chapter 2, the private/public keys are mathematically related and the public key can be generated from the private key any time (deterministic). So, it is not really a requirement to save the public keys. as such. True randomness is not possible through software implementations, so many servers or applications use hardware security modules (HSMs) to generate true random bits and also to protect the private keys. Unlike public keys, private keys definitely require saving them with maximum security. If you lose them, you cannot generate a signature that would justify the ownership of the public key (or Bitcoin address) that received some amount in any transaction. The public keys are hashed twice to generate the Bitcoin adress, first with SHA256 and then with RIPEMD160. This is also deterministic, so given a public key, it is just a matter of a couple of hashes to generate the Bitcoin address.

Note carefully that the Bitcoin address does not really reveal the public key. This is because the addresses are double-hashed public keys and it's quite infeasible to find the public key given the Bitcoin address. However, for someone with a public key, it is easy to claim the ownership of a Bitcoin address. The hashing technique in Bitcoin shortens and obfuscates the public key. While it makes the manual transcription easier, it also provides security against unanticipated problems that might allow reconstruction of private keys from public keys. This is possibly the safest implementation! Public keys are revealed only when the transaction output is being claimed by the owner, not when it was transacted to them, as you can see in Figure 3-28.

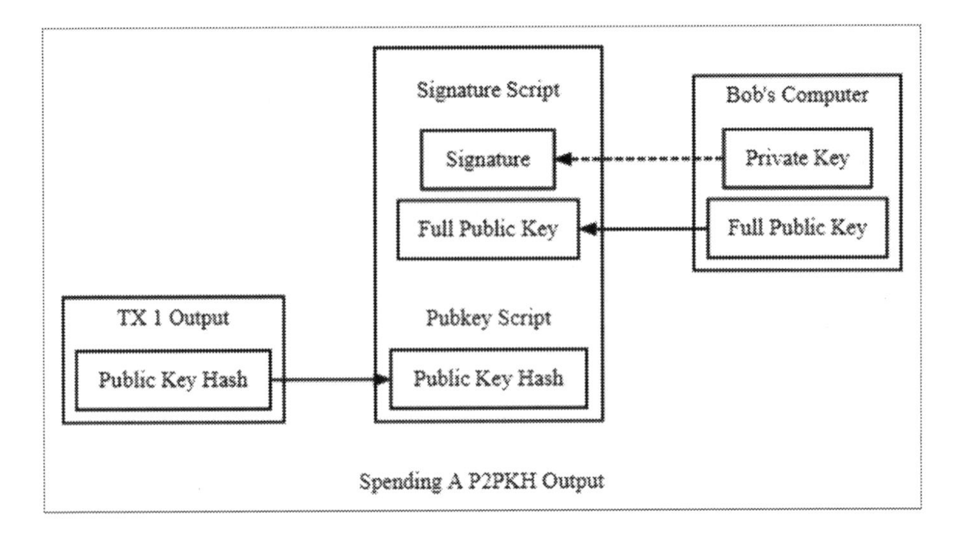

Figure 3-28. *Revealing public key to claim a transaction*

Bitcoin wallets are nothing but the SPVs and are served by the full nodes. We already looked at the functioning of SPVs, so in this section we will take a look at some wallet-specific activities. We all understand that to make a transaction, or to receive a transaction, you need not be running a full node. All you want is a wallet to be able to save your private/public key pair, to be able to make and receive transactions (actually view and verify the ones made to you). We already learned the verification part while going through the SPVs section. Let us take a look at how to initiate a transaction using a wallet.

It is advisible that you run your own full node and connect your wallet to it, as it would be the most secured way of working on Bitcoin. However, it is not a mandate and you can still work without maintaining your own node. Keep in mind that when you query a node, you have to mention your public address to get the list of UTXOs, and the full node is becoming aware of your public address, which is a privacy leak! All a wallet has to do is get the list of UTXOs so it can spend a transaction by signing it with its private key and publish that transaction into the Bitcoin network. This can be done by creating your own wallet software or by using a third-party wallet service. However, be careful with the wallet service providers because you are allowing them to take control of your private key. Whether they deliberately take your Bitcoins or they themselves are hacked, which has been the case with many wallet services, you lose your Bitcoins. At the end of the day, all wallet-service providers are centralized, though the Bitcoin network is decentralized. A typical pictorial representation of initiating a Bitcoin transaction through the wallet software can be represented as shown in the following (Figure 3-29).

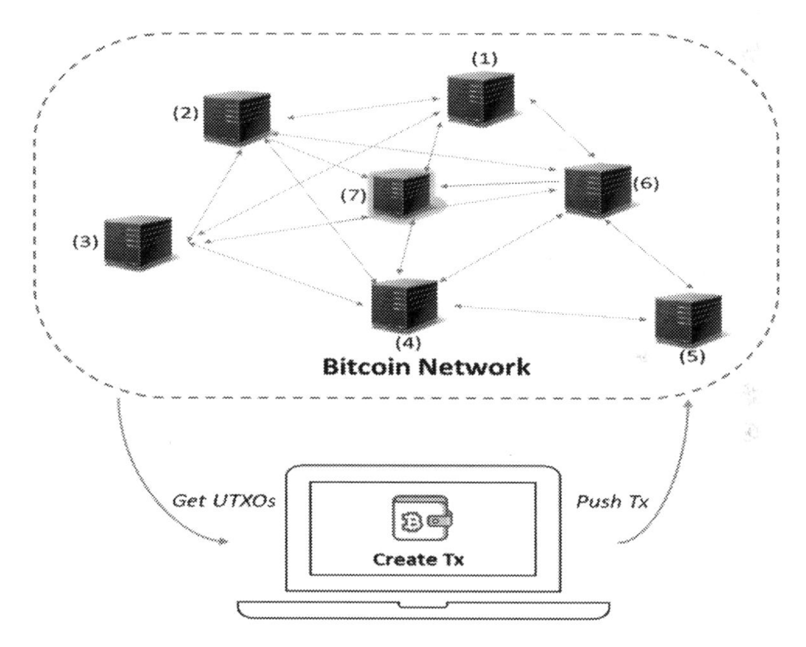

Figure 3-29. *A wallet application interacting with the Bitcoin network*

An example of an SPV client that can serve as a Bitcoin wallet is "BitcoinJ." BitcoinJ is actually a library to work with the Bitcoin protocol, maintain a wallet, and initiate/validate transactions. It does not require a full node such as a Bitcoin Core node locally and can function as a thin client node. Though it is implemented in Java, it can be used from any JVM-compatible language such as JavaScript and Python.

Summary

In this chapter, we learned how blockchain concepts we discussed in the previous chapter were put together to build Bitcoin as a cryptocurrency use case of blockchain technology. We covered the evolution of Bitcoin, the history of it, what it is, the design benefits, and why it is so important. We got to know about granular details on the Bitcoin network, transactions, blocks, the blockchain, consensus, and how all these are stitched together. Then we learned about the requirement of a wallet solution to interact with the Bitcoin blockchain system.

In the 1990s, mass adoption of the Internet changed the way people did business. It removed friction from creation and distribution of information. This paved the way for new markets, more opportunities, and possibilities. Similarly, blockchain is here today to take the Internet to a whole new level. Bitcoin is just one cryptocurrency application of blockchain, and the possibilities are limitless. In the next chapter, we will learn about how Ethereum works and how it has become a defacto standard for various decentralized applications on one public blockchain network.

References

Bitcoin: A Peer-to-Peer Electronic Cash System

Nakamoto, Satoshi, "Bitcoin: A Peer-to-Peer Electronic Cash System," `https://bitcoin.org/bitcoin.pdf`.

All about Bitcoin Network and Transactions

Bitcoin wiki, `https://en.bitcoin.it/`.

Blockchain Technology

Crosby, Michael, Nachiappan; Pattanayak, Pradhan, Verma, Sanjeev, Kalyanaraman, Vignesh, "BlockChain Technology: Beyond Bitcoin," Sutardja Center for Entrepreneurship & Technology, University of California, Berkeley, `http://scet.berkeley.edu/wp-content/uploads/BlockchainPaper.pdf`, October 16, 2015.

Accelerating Bitcoin's Transaction Processing

Sompolinsky, Yonatan, Zohar, Aviv, "Secure High-Rate Transaction Processing inBitcoin," Hebrew University of Jerusalem, Israel, School of Engineering and Computer Science, `https://eprint.iacr.org/2013/881.pdf`.

CHAPTER 4

How Ethereum Works

The era of blockchain applications has just begun. Ethereum is here to be the defacto blockchain platform for building decentralized applications. We already learned in the previous chapters that public blockchain use cases are not just limited to cryptocurrencies, and the possibilities are only limited by your imagination! Ethereum has already made inroads in many business sectors and works best not only for public blockchain use cases, but also for the private ones. Ethereum has already set a benchmark for blockchain platforms and must be studied well to be able to envision how usable decentralized applications can be built with or without using Ethereum. Today, it is possible to build blockchain applications with minimal knowledge of cryptography, game theory, mathematics or complex coding, and computer science fundamentals, thanks to Ethereum.

In Chapter 3, we learned how Bitcoin works by taking a deep dive into the protocol as well as the Bitcoin application. We witnessed how the cryptocurrency aspect is so much interwoven into the Bitcoin protocol. We learned that Bitcoin is not Bitcoin on blockchain, rather a Bitcoin blockchain. In this chapter, we will learn how Ethereum has successfully built an abstract foundation layer that is capable of empowering various different blockchain use cases on the same blockchain platform.

B. Singhal et al., *Beginning Blockchain*, https://doi.org/10.1007/978-1-4842-3444-0_4

From Bitcoin to Ethereum

Obviously, blockchain technology came along with Bitcoin back in 2009. After Bitcoin stood the test of time, people believed in the potential of blockchain. The use cases now have gone beyond banking and finance sectors and have enveloped other industries such as supply chain, retail, e-commerce, healthcare, energy, and government sectors as well. This is because different flavors of blockchain have come up and address specific business problems. Nonetheless, there are public blockchain platforms such as Ethereum that allow different decentralized use cases to be built on the same public Ethereum platform.

With Bitcoins, decentralized peer-to-peer transaction of cryptocurrency was possible. People realized that blockchain could be used to transact and keep track of anything of value, not just cryptocurrency. People started exploring if the same Bitcoin network could be used for any other use case. To give you an example, "proof of existence" is one such use case where the hash of a document was injected in the Bitcoin blockchain network so that anyone could later verify that such a ducument was existant in so and so point in time. Vitalik Buterin introduced the Ethereum blockchain platform that could facilitate transactions of not just money, but also shares, lands, digital content, vehicles, and many others that have some intrinsic value. Take a look at Figure 4-1.

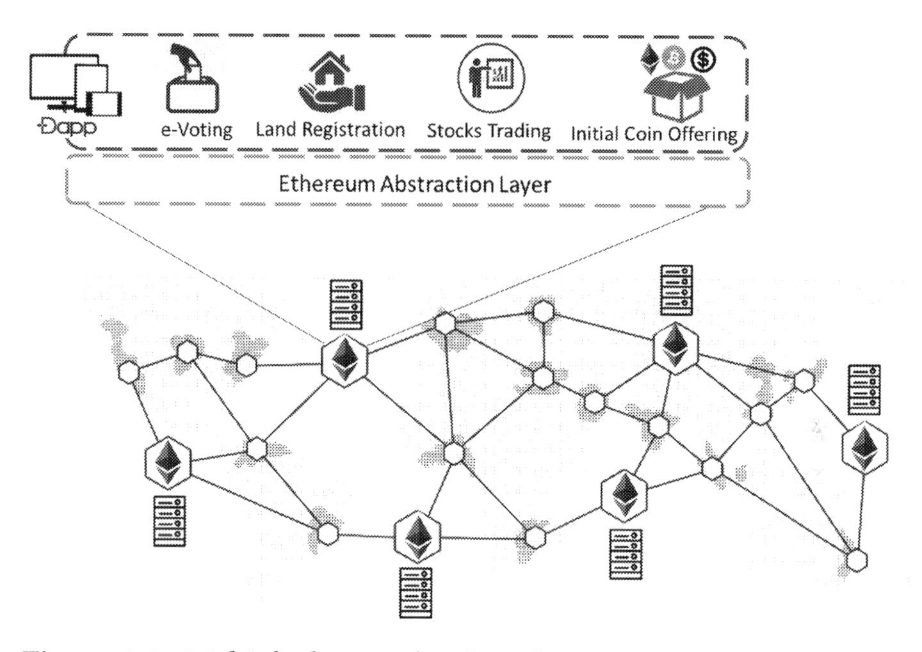

Figure 4-1. *Multiple decentralized applications on one Ethereum platform*

Like Bitcoin, Ethereum is a public blockchain platform with a different design philosophy. The most innovative approach was to build an abstraction layer so that transactions from different applications are generalized to the program code that can run on all the Ethereum nodes. Even in Ethereum, the miners generate Ether, a tradeable cryptocurrency because of which the public blockchain network is self-sustainable. Any application that is running on Ethereum has to pay transaction fees that eventually the miners get for running the nodes and sustaining the whole network.

Ethereum as a Next-Gen Blockchain

With the Bitcoin blockchain, the developer community tried building different decentralized applications with a completely new blockchain, or were trying to modify Bitcoin Core to increase the set of functionalities.

Either way, it was complicated as well as time consuming. A different design with an alternative protocol was probably the need of the hour then, which is why the Ethereum blockchain platform! The purpose was to facilitate development of many blockchain applications on one Ethereum platform rather than building dedicated blockchains for each application separately. Ethereum enabled rapid development of decentralized applications that could interact among themselves, ensuring adequate security. As mentioned in the previous section, Ethereum does this by building an abstract foundation layer. Unlike Bitcoin, Ethereum supported Turing-complete language so anyone could write smart contracts that could virtually do anything and everything on a programming perspective. Also, Ethereum is stateful by design and keeps track of the acount states, which is very different from Bitcoin where everything remains as a transaction and there is no internal persistent memory for scripts. With the help of an abstract foundation layer, the underlying complexities are hidden from the developers and not just that; the developers get the flexibility of designing their own state transition functions for direct transfer of value and information, and transaction formats.

In an effort to meet the objective, the core innovation of Ethereum was the Ethereum Virtual Machine (EVM). The support for Turing-complete languages through the EVM makes it easy for the developers to create blockchain applications. Just the way a Java Virtual Machine (JVM) is required to run Java code, EVM is required to run the smart contracts. For now, just keep in mind that smart contracts are the Ethereum scripts written in a Turing-complete language that automatically gets executed in case a predefined event occurs. The "ScriptSig" and "ScriptPubKey" in Bitcoins are the basic versions of smart contracts so to speak. We learned in the previous chapter that in Bitcoins, the instruction set was very limited. In Ethereum, however, one could code almost any program that would run on the EVM on each and every node in the Ethereum blockchain network. The decentralized applications in Ethereum are called DApps. Ethereum being a global decentralized computer system with no centralized server, DApps are the applications that run without downtime, fraud, or any sort of regulations. A peer-to-peer electronic cash system such as Bitcoin is very easy to build on Ethereum as a DApp. Similarly, any other asset with some intrinsic value, such as land, cars, houses, votes, etc., could easily be transacted through their respective DAaps on Ethereum in the form of tokens.

Unlike traditional software development and deployment, DApps do not need to be hosted on a back-end server. The "code" is embedded as payload in transactions, so to speak, that are then sent to the mining nodes in the Ethereum network. Such transactions would be considered by the mining ecosystem because of the ETH (Ether) paid as "*gas* Price." Like in Bitcoin, these transactions get broadcast to other miners in the network that they are accessible to. The transaction then eventually gets into a block and becomes an eternal part of the blockchain when consensus is reached. Developers have the liberty to code up any solution and deploy that in the Ethereum network. The network executes that, all by itself, and validates and produces the outputs as well. Well, had it been without any cost, the network wouldn't have been sustainable. There is a *gas* Price associated with each blockchain transaction, and writing some garbage code and deploying that into the Ethereum network could be an expensive affair!

Design Philosophy of Ethereum

Ethereum borrows many concepts from Bitcoin Core as it stood the test of time, but is designed with a different philosophy. Ethereum development has been done following certain principles as follows:

- **Simplistic design**: The Ethereum blockchain is designed to be as simple as possible so that it is easy to understand and develop decentralized applications on. The complexities in the implementation are kept to a bare minimum at the consensus level and are managed at a level above it. As a result, high-level language compilation or serialization/deserialization of arguments, etc. are not a concern for the developers.

- **Freedom of development**: The Ethereum platform is designed to encourage any sort of decentralization on its blockchain platform and does not discremenate or favor any specific kinds of use cases. This freedom is given to an extent that a developer can code up an infinite loop in a smart contract and deploy it. Obviously, the loop will run as long as they are paying the transaction fee (*gas* Price), and the loop eventually terminates when it runs out of *gas*.
- **No notion of features**: In an effort to make the system more generalized, Ethereum does not have built-in features for the developers to use. Instead, Ethereum provides support for Turing-complete language and lets the users develop their own features the way they want to. Starting from basic features such as "locktime," as in Bitcoin till full blown use cases, everything can be coded up in Ethereum.

Enter the Ethereum Blockchain

We learned about the objective behind Ethereum blockchain and its design philosophy. To be able to understand and appreciate this next-gen blockchain and build decentralized applications on it, we will learn about the core components of Ethereum in great detail in this section.

Ethereum Blockchain

The Ethereum blockchain data structure is pretty similar to that of Bitcoin's, except that there is a lot more information contained in the block header to make it more robust and help maintain the state properly. We will learn more about the Ethereum states in the following sections. Let us focus more on the blockchain data structure and the header in this section. In Bitcoins, there was only one Merkle root in the block header for all the transactions in a block. In Ethereum, there are two more Merkle roots, so there are three Merkle roots in total as follows:

- **stateRoot**: It helps maintain the global state.
- **transactionsRoot**: It is to track and ensure integrity of all the transactions in a block, similar to Bitcoin's Merkle root.
- **receiptsRoot**: It is the root hash of the receipts *trie* corresponding to the transactions in a block

We will take a look at these Merkle roots in their respective sections of block header information. For better comprehension, take a look at Figure 4-2.

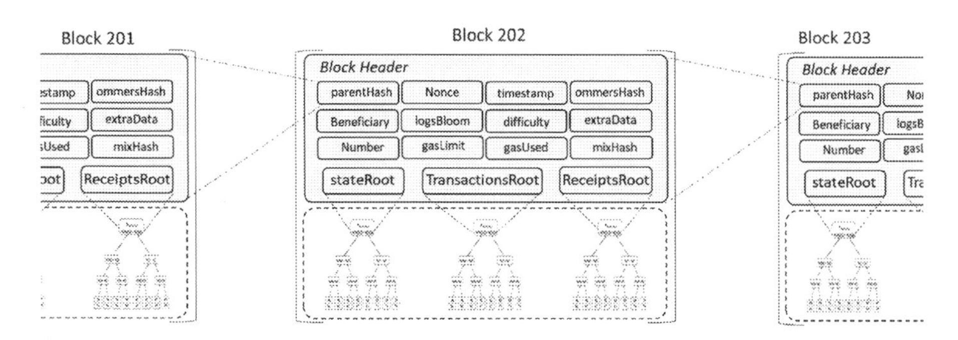

Figure 4-2. *The blockchain data structure of Ethereum*

Every block usually comprises block header, transactions list, uncles list, and optional extraData. Let us now take a look at the header fields to understand what they mean and their purpose for being in the header. While you do so, keep in mind that there could be slight variants of these names in different places, or the order in which they are presenbted could be different in different places. We suggest that you build a proper understanding of these fields so that any different terminology that you might come across won't bother you much.

Section-1: Block metadata

- parentHash: Keccak 256-bit hash of the parent block's header, like that of Bitcoin's style
- timestamp: The Unix timestamp current block
- number: Block number of the current block
- Beneficiary: The 160-bit address of "author" account responsible for creating the current block to which all the fees from successfully mining a block are collected

Section-2: Data references

- transactionsRoot: The Keccak 256-bit root hash (Merkle root) of the transactions trie populated with all the transactions in this block
- ommersHash: It is otherwise known as "uncleHash." It is the hash of the uncles segment of the block, i.e., Keccak 256-bit hash of the ommers list portion of this block (blocks that are known to have a parent equal to the present block's parent's parent).
- extraData: Arbitrary byte array containing data relevant to this block. The size of this data is limited to 32 bytes (256-bits). As of this writing, there is a possibility that

this field might become “extraDataHash”, which will point to the “extraData” contained inside the block. extraData could be raw data, charged at the same amount of *gas* as that of transaction data.

Section-3: Transaction execution information

- stateRoot: The Keccak 256-bit root hash (Merkle root) of the final state after validating and executing all transactions of this block
- receiptsRoot: The Keccak 256-bit root hash (Merkle root) of the receipts trie populated with the recipients of each transaction in this block
- logBloom: The accumulated Bloom filter for each of the transactions' receipts' Blooms, i.e., the “OR” of all of the Blooms for the transactions in the block
- *gasUsed*: The total amount of *gas* used through each of the transactions in this block
- *gasLimit*: The maximum amount of *gas* that this block may utilise (dynamic value depending on the activity in the network)

Section-4: Consensus-subsystem information

- difficulty: The difficulty limit for this block calculated from the previous block's difficulty and timestamp
- mixHash: The 256-bits mix hash combined with the 'nonce' for the PoW of this block
- nonce: The nonce is a 64-bit hash that is combined with mixHash and can be used as a PoW verification.

Ethereum Accounts

The Ethereum accounts, unlike Bitcoins, are not in the form of unspent transaction outputs (UTXOs). In the Bitcoin chapter, we learned that Bitcoins are actually present in the form of transactions that have an owner (owner's public key, 20-byte address) and a value. The owner can spend the transaction if they have the valid private key for the transaction they are trying to spend. Bitcoin therefore is a state transition system where "state" refers to the collection of all UTXOs. Every time a block is mined, a state change happens because each block contains a bunch of transactions where each transaction cosumes UTXO(s) and produces UTXO(s). Note here that the state is not encoded inside the blocks. So, there is no notion of an account balance as such in Bitcoin's design. Ethereum on the other hand is stateful, and its basic unit is the account. Each account has a state associated with it and also has a 20-byte (160 bits) address through which it gets identified and referenced. The purpose of blockchain in Ethereum is to keep track of the state changes. There are broadly two types of Ethereum accounts:

- **Externally Owned Accounts (EOAs)**: These accounts are also known as "simple accounts" that are usually owned by users or devices who control these accounts using Private Keys. The EOAs can send transactions to other EOAs or Contract Accounts by signing with a private key. The transaction between two EOAs is usually to transfer any form of value. On the other hand, when an EOA makes a transaction to a Contract Account, the purpose is to activate the "code" inside the Contract Account.
- **Contract Accounts**: These are controlled only by the code contained in them. This code inside the Contract Accounts is referred to as "smart contracts."

They are usually activated when a transaction is sent to the Contract Account by the EOAs or by other Contract Accounts. Even though the Contract Accounts are capable of executing complex business logics through the code they contain, they can't initiate new transactions on their own and always depend on the EOAs. All they can do is respond to other transactions (obviously by making transactions) as per the logic coded in their "code."

Take a look at the following three scenarios (Figures 4-3 to 4-5) to get a better understanding on the communication between the EOAs and Contract Accounts.

EOA to EOA transaction:

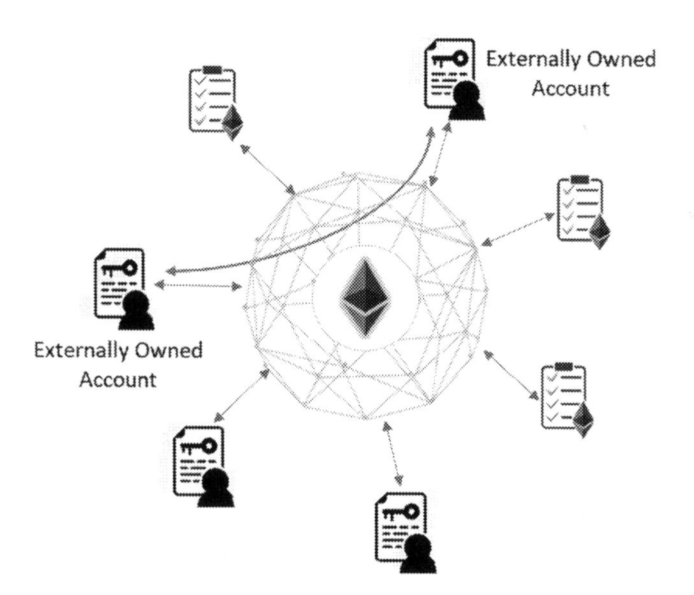

Figure 4-3. *EOA to EOA transaction*

EOA to Contract Account Transaction:

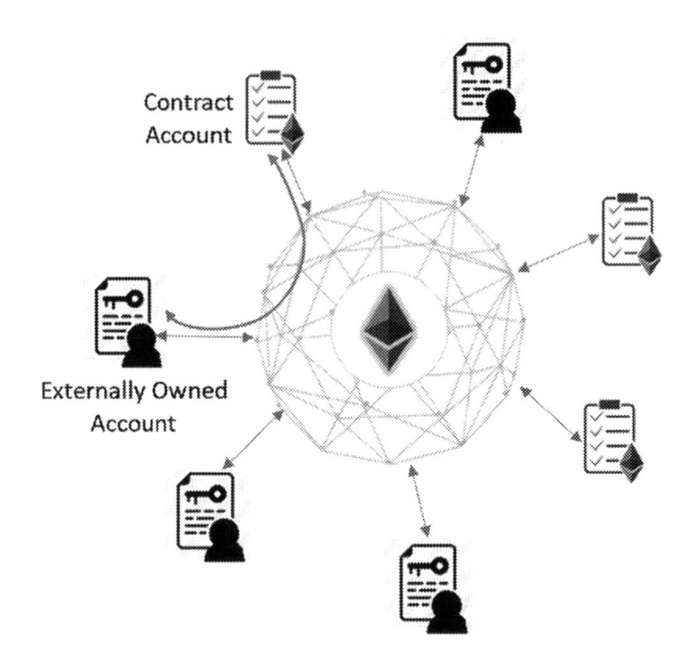

Figure 4-4. *EOA to Contract Account transaction*

EOA to Contract Account to other Contract Account transaction:

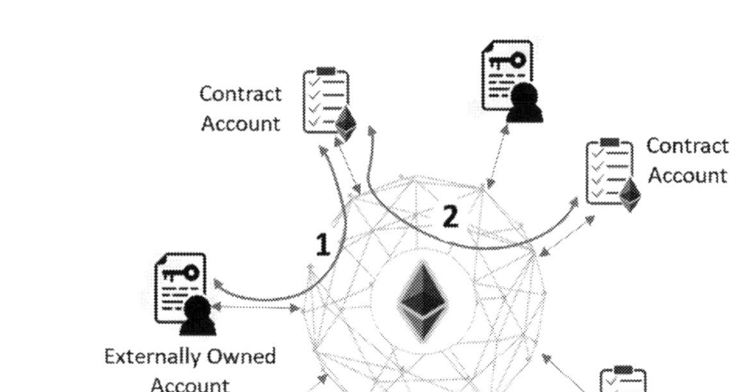

Figure 4-5. *EOA to Contract Account to Contract Account transaction*

Just so the previous representations are not confusing, please be aware that the Contract Accounts are internal and the communications between them, too. Unlike EOA accounts where EOAs make a transaction that gets injected in the blockchain, Contract Accounts and the transactions between them are internal phenomena.

Advantages of UTXOs

We must understand that Bitcoin's design perspective was to maintain anonymity to an extent possible. When we compare it with Ethereum, the following advantages of UTXOs seem to have a lot of significance:

- **Better privacy**: In Bitcoins, it is advisible to use a new address while receiving transactions, which helps reinforce anonymity. Even with sophisticated statistical or machine learning techniques, it is difficult to link the accounts together, though not impossible.

- **Potentially more scalable**: The discussion pertaining to scalability is usually very subjective and depends on the context, use case at hand, and many other factors. The intention here is to just mention UTXO's inherent potential to scale. It is very easy to execute the transactions in parallel. Also, when an owner or other nodes maintaining the Merkle proof of ownership data for some coins lose this data, only the owner is impacted. On the contrary, when Merkle tree data for some account is lost, then any operation on that account would not be feasible, even sending to it.

Advantages of Accounts

Even though Ethereum in a way is an extention to Bitcoin, it is imagined with a whole new design with its own set of pros-cons tradeoff. Let us take a look at the following advantages of Ethereum accounts compared with Bitcoin design:

- **Significant space saving**: In Bitcoins, when multiple transactions are clubbed together to make one transaction (e.g., if you have to make a 5BTC transaction and you never received one transaction with at least 5BTC that you could use in this case, then you have to bundle multiple transactions so the total exceeds 5BTC), that many references to those individual transactions must be made. Also, all those transactions must have different addresses, so as many transactions, that many addresses also! In Ethereum accounts, however, just one reference to an account is good enough. Even though Ethereum uses Merkle

Patricia tree (MPT), which is a bit more space intensive than Merkle tree, you end up saving a significant amount of space for complex transactions.

- **Simple to code**: Along with UTXOs and scripts that are not Turing-complete, it is difficult to design complex systems. UTXOs can either be spent or unspent; there is no other state possible in between. Which makes it difficult to code up complex business logics. Even if the scripts are empowered to do more, it gets more complicated as compared with just using accounts. Since the objective of Ethereum is to go beyond cryptocurrency and accommodate different kinds of use cases (through DApps), an accounts-based system is almost inevitable.
- **Lightweight client reference**: Unlike Bitcoin clients, Ethereum client applications can easily and quickly access all the data related to an account by scanning down the state tree in a specific direction. In the UTXO model, there are usually multiple references to multiple transactions associated to any specific transaction under consideration.

Account State

We learned that every account has a state associated with it. We also looked at the two kinds of accounts that exist with Ethereum, one is a Contract Account and the other is an Externally Owned Account or EOA. Regardless of the account type, they are tracked by the "stateRoot" Merkle root in the block header and may appear as shown in Figure 4-6.

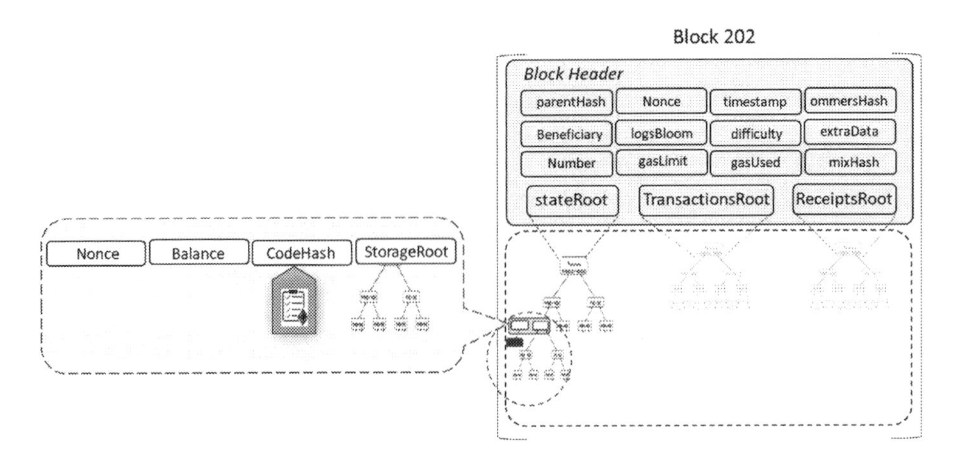

***Figure 4-6.** Zooming in to account state representation*

As you can see in the figure, irrespective of whether the account is an EOA or or a Contract Account, it has the following four components:

- **Account balance**: Total "Ether" balance in the account. More precisely, number of *Wei* owned by the address (1ETH = 10^{18} Wei)
- **CodeHash**: This is the hash of the "code." Every Contract Account has "code" in it that gets executed on the EVM. The hash of this code is stored in this CodeHash field. For the EOA accounts, however, there is no "code," so the CodeHash field contains the hash of empty string.
- **StorageRoot**: It is the 256-bit root hash of Merkle tree that encodes the storage contents of an account. The MPT encodes the hash of the storage content. Keeping the root hash of this tree in the StorageRoot field helps track the content of an account and also helps ensure its integrity.

- **Nonce**: It is a counter that ensures each transaction is processed only once. For EOAs, this number represents the number of transactions from the account's address. For Contract Accounts, it represents the number of contracts created by this account.

So, it is the "state" trie that is responsible to keep track of the state changes of Ethereum blockchain. However, what is a bit tricky is that the state is not directly stored in each block, rather in the form of Recursive Length Prefix (RLP)-encoded state data in MPT at every Ethereum node. So, to maintain the global state, the Ethereum blockchain includes "state roots" in each and every block that store the root hash of the hash tree (Merkle root) representing the system state at the time the block was created.

As per the Ethereum Yellow Paper, the "World State" is a mapping between addresses (160-bit identifiers) and account states. So, the World State has the information of all the accounts in blockchain, but is not stored in each block. Each block only modifies parts of the state. In a way, the World State is generated processing each block since the genesis block. Certain Ethereum nodes can choose to maintain all historical states by keeping all the historical transactions, that is, state transitions and their outputs. This allows clients to query the state of the blockchain at any time, even for the historic ones, without having to recalculate everything from the beginning. Retrieving the state information is similar to an aggregate query in SQL where data is readily available; just aggregation is required. So, old state data can easily be discarded (this is known as "pruning") because they can be computed back when required. Well, the state data by design is *implicit dada*, which means state information should only be calculated.

Trie Usage

We learned the three types of tries that have their roots in the block header. These roots are basically the pointers to those three tries. Though we looked at the one-liner explanations of these tries in previous sections, let us just revisit them with a slightly different choice of words

- State trie: It represents the entire state (the global state) after accessing the block.
- Transaction trie: It represents all the transactions in a block keyed by index (i.e., key:0 for the first transaction to execute, key:1 for the second transaction, etc.). Recollect the MPT fundamentals we covered earlier and try to correlate.
- Receipt trie: It represents the "receipts" corresponding to each transaction. A receipt for a transaction is an RLP-encoded data structure as shown following:

```
[ medstate, gas_used, logbloom, logs ]
```

Let's now dig deeper into the Receipt trie as we havn't covered the basics yet on this. Take a look at all the fields in the Receipt trie's RLP-encoded data structure and follow through the following descriptions for those fields:

- medstate: It is the State trie root after processing the transaction. A successful transaction updates the Ethereum state.
- *gas_used*: It is the total amount of *gas* used for processing the transaction.
- logs: It is a list of items of the form-

```
[address, [topic1, topic2...], data]
```

- These list items are produced by the LOG0, LOG1... *opcodes* during the execution of the transaction. The "address" field is the address of the contract that produced the log, the "topic" fields are up to four 32-byte values, and the "data" field is an arbitrarily sized byte array.
- Logbloom: It is a Bloom filter made up of the addresses and topics of all logs in the transaction. This is different from the one present in the block header.

Merkle Patricia Tree

In Ethereum, the accounts are mapped with their respective states. The mapping between all the Ethereum accounts, including EOAs and Contract Accounts with their states, is collectively referred to as World States. To store this mapping data, the datastructure used in Ethereum is the MPT. So, MPT is the principal data structure used in Ethereum which is otherwise known as Merkle Patricia trie. We learned about the Merkle trees in the Bitcoin chapter, which already takes us half way through in understanding MPT. MPT is actually derived by taking elements from both Merkle tree and Patricia tree.

Recollect from the Bitcoin chapter that Merkle trees are the binary hash trees where the leaf nodes contain the hash of the data blocks and every nonleaf node contains the hashes of their child nodes. When such a data structure is implemented, it becomes easy to check if a certain transaction was a part of a block. Only by using very little information from the entire block, that is, by using just the Merkle branch instead of the entire tree, providing proof of membership was quite easy. Merkle trees facilitate efficient and secure verification of the contents in decentralized systems. Instead of downloading every transaction and every block, the light clients can only download the chain of block headers, that is, 80-byte chunks of data for each block that contain only five things: hash of the previous block header, timestamp, mining difficulty, nonce value that satisfied PoW, and the root hash of the Merkle tree containing all the transactions for that block. While it is quite useful and interesting, note here that apart from validating the proof of membership for a transaction in a block, there is nothing much you could do. One particular limitation is that no information can be proved about the current state (e.g., total digital asset holdings, name registrations, status of financial contracts). Even to check how many Bitcoins you hold, quite a lot of querying and validating is involved.

Patricia trees on the other hand are a form of Radix trees. The name PATRICIA stands for "Practical Algorithm to Retrieve Information Coded In Alphanumeric." A Patricia tree facilitates efficient insert/delete operations. The key-value lookups in the Patricia tree are very efficient. Keys are always encoded in the path. So, "key" is the path that you take from the root till the leaf node where the "value" is stored. Keys are usually the strings that help descend down the path where each character indicates which child node to follow to reach the leaf node and find the value stored in it.

So, the MPTs provide a cryptographically authenticated data structure used to store all (key, value) bindings in Ethereum. They are fully deterministic, meaning that a Patricia tree with the same (key, value) bindings will surely be the same down to the last byte. The insert, lookup, and delete operations are quite efficient with O(log(n)) complexity. Due to the Merkle part in MPT, hash of a node is used as the pointer to the node and the MPT is constructed accordingly, where

Key == SHA3(RLP(value))

While the Merkle part provides a tamperproof and deterministic tree structure, the Patricia part provides an efficient information retrieval feature. So, if you notice carefully, the root node in MPT becomes a cryptographic fingerprint of the entire data structure. In the Ethereum P2P network, when transactions are broadcast over the wire, they are assembled by every mining node that received them. The nodes then form a Tree (a.k.a. trie) and compute the root hash to include in the Block header. While the transactions are stored locally in the tree, they are sent to other nodes or clients after they are serialized to lists. The receiving parties have to deserialize them back to form the transaction tree to verify against the root hash. Also note that in Ethereum, MPTs are a little modified for better fitment with Ethereum implementation. Instead of binary, hexadecimal is used–X characters from a 16 character "alphabet." Hence nodes in the tree or trie have 16 child nodes (the 16 character hex alphabet) and a maximum depth of X. Just to let you know, a hex character is referred to as a "nibble" in many places.

The basic idea of an MPT in Ethereum is that for a single operation, it will only modify the minimum amount of nodes to recalculate the root hash. This way the storage and complexities are kept minimal.

RLP Encoding

You must have noticed that we mentioned RLP encoding in previous sections. We will give you a heads-up on what it is all about in this section. RLP stands for Recursive Length Prefix. It is a serialization method used in Ethereum for blocks, transactions, and wire protocol messages while sending data over the wire and also for account state data while saving the state in Patricia tree. In general, when complex data structures need to be stored or transmitted and then get reconstructed at the receiving end for processing, object serialization is a good practice. RLP in that sense is similar to JSON and XML, but RLP is believed to be more minimalistic, space efficient, simple to implement, and guarantees absolute byte-perfect consistency. This is why RLP was chosen to be the main serialization technique for Ethereum. Its sole purpose is to store nested arrays of raw bytes. It does not try to define any specific data types either, such as Booleans, floats, doubles, integers, etc., and is only designed to store structure in the form of nested arrays. Key/value maps are not explicitly supported by RLP. So, it is advisible to represent such maps as [[k1, v1], [k2, v2], ...], where k1, k2... are in lexicographic order (sorted using the standard ordering for strings). Alternatively, use the higher-level Patricia tree encoding that has an inherent RLP encoding scheme.

Please keep in mind that RLP is used only to encode the structure of the data and is completely unaware of the type of object being encoded. While it helps reduce the size of the encoding array of raw bytes, the decoding end must be aware of the type of object it is trying to decode.

Ethereum Transaction and Message Structure

In the previous section, we looked at the block structure and the different fields in the block's header. For a transaction to be qualified by the miners or Ethereum nodes, it has to have a standardized structure. A typical Ethereum **transaction** (e.g., what you pass through sendRawTransaction() that we will see later in this book) consists of the following fields:

- nonce: It is an integer, just a counter equal to the number of transactions sent by the sender account, i.e., transaction sequence number.
- *gasPrice*: Price you are willing to pay in terms of the number of Wei to be paid per unit of *gas*
- *gasLimit*: The maximum amount of *gas* that should be used in executing this transaction, which also limits the maximum number of computational steps the transaction execution is allowed to take
- To: Recipient's 160-bits address or Contract's address. For the transaction that is used to create a contract (it means contract's address does not exist yet), it is kept empty.

- Value: Total Ether (number of Wei) to be transferred to the recipient by the transaction sender
- *V, r, s*: values corresponding to the ECDSA signature of the transaction; also represent the sender of this transaction
- init: This is not really an optional field, only used with transactions used for creating contracts. This field can contain an unlimited size byte array specifying the EVM-code for the account initialisation procedure.
- The opcode "init" is used only once for initializing the new Contract Account and gets discarded after that. It returns the body of the account code after associating it with the Contract Account. Keep in mind that this association is a permanent phenomenon and never changes.
- Data: An optional field that can contain a message to be sent to a contract or simple account. It has no special function as such by default, but the EVM has an *opcode* —using which, a contract can access this data field and perform necessary computations and place them in storage.

Note carefully that the aforementioned fields are supplied in the order specified and are all RLP encoded, except for the field names. So, an Ethereum transaction actually means a signed data package with these fields. The *gasPrice* and *gasLimit* fields are important to prevent denial of service attack. In order to prevent accidental or deliberate attempts of infinite loops or other computational wastage in code, each transaction is required to set a limit on how many computational steps for code execution it can use.

Ethereum transactions are actually the "state transition functions" because a successful transaction changes the state. Also, the result of these transactions can be stored, as we already looked at in the "Account State" section previously.

Ethereum **messages** on the other hand are like transactions, but are triggered only by Contract Accounts and not by EOAs. Also, messages are only meant to be between the Contract Accounts, due to which they are also referred to as "internal transactions." So, contracts have the ability to send messages to other contracts.

Typically, a message is produced when a contract, while executing its code, encounters the "CALL" or "DELEGATECALL" opcodes. So, messages are more like function calls that exist in the Ethereum execution environment. It is also important to note that messages are always raw and never serialized or deserialized. A message contains the following fields:

- Sender: The sender of the message as an implicit option
- Recipient: The recipient contract address to send to
- Value: The amount of Wei to transfer to the contract address along with the message
- Data: Optional field, but can contain input data for the recipient contract provided by the sender
- *gasLimit*: The value that limits the maximum amount of *gas* the code execution can consume when triggered by the message. It is also termed "*startGas*."

We looked at the transaction and messages. An Ethereum transaction can be from an EOA to an EOA, or from an EOA to a Contract Account. There exists another situation where a transaction from an EOA is initiated to create a Contract Account (recollect the "init" field that we just covered).

Now, just think about what exactly a transaction is? It is definitely the bridge between the external world and the Ethereum blockchain, but what more? If you zoom in to a transaction, you will see that it is an instruction, initiated by the EOA by signing it, which gets serialized and submitted to the blockchain. Take a look at Figure 4-7.

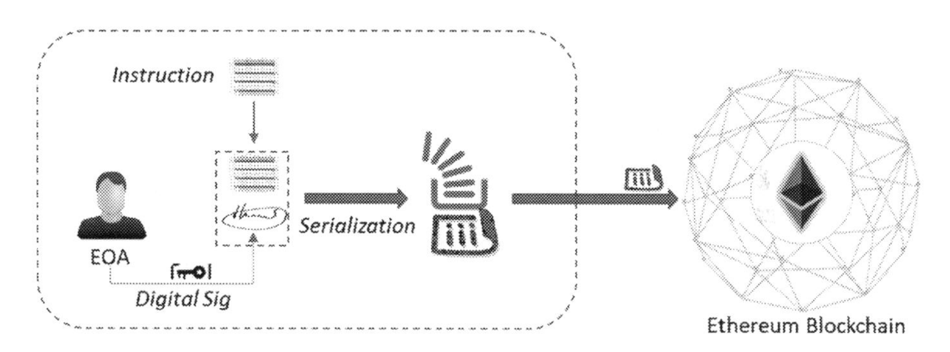

Figure 4-7. *Transaction initiation–zoomed in*

Now what happens after a transaction is injected into the blockchain? Well, it starts executing at every Ethereum node if found valid. While this transaction is undergoing execution, Ethereum is designed to keep tabs on the "substate" to track the flow of execution. This is because, if a transaction does not complete due to "running out of *gas*," then the entire execution so far has to be reverted. Also, information collected during the execution is required immediately after the transaction completion. So, the substate contains the following:

- Self-destruct set: a set of accounts (if any) that will be discarded after the transaction completion
- Log series: archived and indexable "checkpoints" of the EVM's code execution to track the contract calls

- Refund balance: It is the amount to be refunded to the sender account post transaction execution. Storage in Ethereum is quite expensive, so there is an SSTORE instruction in Ethereum that is used as a refund counter. The refund counter starts at zero (no refund state) and gets incremented every time the transaction or contract deletes something from the storage. Please note that this refund amount is different and in addition to the unused *gas* that gets refunded to the sender.

In the earlier versions of Ethereum, whether a transaction or contract executes successfully or fails in between, the entire *gas* used to get consumed. This was not always making sense. If an execution stopped due to some authorization/permission issue or any other issue, the execution would stop and the remaining *gas* would still be consumed. The last Byzantium update introduced the "revert" code like an exception handling. In case a contract has to stop, "revert" could be used to revert state changes, return a reason for failure, and credit the remaining *gas* back to the sender. Post successful execution of the transactions or contracts, a state transition happens that we will dive deeper into in the followiung section.

Just the way we looked at blockchaininfo to see a live Bitcoin transaction, if you take a look at `https://etherscan.io` for Ethereum you will find the following information:

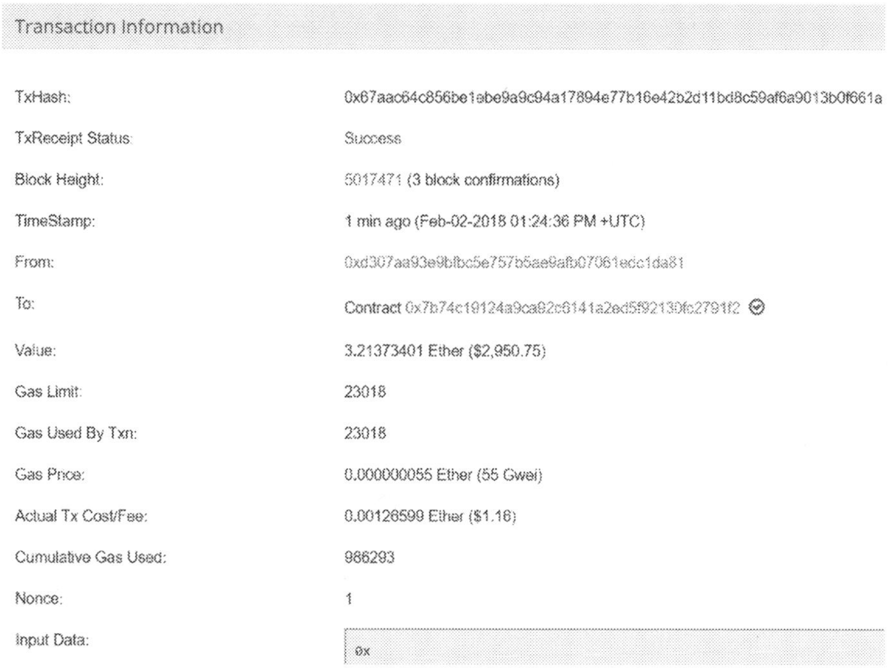

Transaction Information	
TxHash:	0x67aac64c856be1abe9a9c94a17894e77b16e42b2d11bd8c59af6a9013b0f661a
TxReceipt Status:	Success
Block Height:	5017471 (3 block confirmations)
TimeStamp:	1 min ago (Feb-02-2018 01:24:36 PM +UTC)
From:	0xd307aa93e9bfbc5e757b5ae9afb07061edc1da81
To:	Contract 0x7b74c19124a9ca92c6141a2ed5f92130fc2791f2
Value:	3.21373401 Ether ($2,950.75)
Gas Limit:	23018
Gas Used By Txn:	23018
Gas Price:	0.000000055 Ether (55 Gwei)
Actual Tx Cost/Fee:	0.00126599 Ether ($1.16)
Cumulative Gas Used:	986293
Nonce:	1
Input Data:	0x

Ethereum State Transaction Function

In the previous section, we learned about Ethereum transactions and messages. We are now aware that a state transition happens whenever a transaction is through–successfully. So, the state transition function in Ethereum is:

```
APPLY(S,Tx) -> S'    \\where S is old state and S' is the new state
```

Take a look at Figure 4-8.

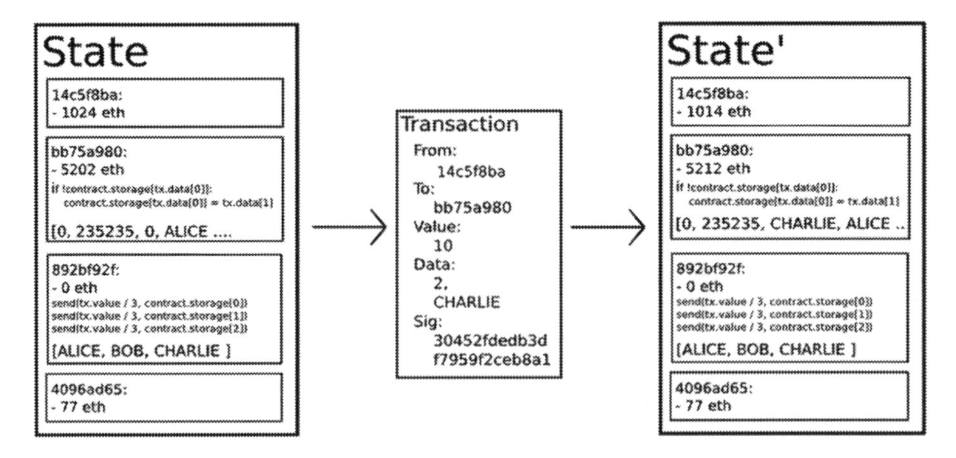

Figure 4-8. *Ethereum state transition function*

So, the state transition function when Tx is applied to state S to result in changed state S′ can be defined as follows:

- Validate the transaction to see if it is well formed.
    - Has the right number of values
    - The signature is valid.
    - The nonce matches the nonce in the sender's account.

  If any of preceding points is not valid, return an **error**.

- Calculate the fee and settle the accounts.
    - Compute the transaction fee as *gasLimit * gasPrice.*
    - Determine the sending address from the signature.
    - Subtract the fee from the sender's account balance and increment the sender's nonce.

If there is not enough balance to spend, return an **error**.

- Initialize GAS = *gasLimit*, and take off a certain quantity of *gas* per byte to pay for the bytes as a transaction fee.
- Transfer the transaction value (could be anything of value) from the sender's account to the receiving account. Note here that the transaction could be for anything of some intrinsic value such as land, vehicle, ERC20 tokens, etc., but the *gas* Price has to be in Ether so that the miners would accept the transaction. If the receiving account does not yet exist, create it.

  If the receiving account is a contract and not an EOA, then run the contract's code either to completion or until the execution runs out of *gas*. Note here that the contract code gets executed on every node's EVM as part of the block validation process so that the block, hence the contract's output post execution, becomes a part of the main blockchain.
- If the value transfer failed because the sender did not have enough money, or the code execution ran out of *gas*, revert all state changes (thanks to MPT implementation) except the payment of the fees, and add the fees to the miner's account.
- Otherwise, refund the fees for all remaining *gas* back to the sender, and send the fees paid already for *gas* consumed to the miner.

Gas and Transaction Cost

Transactions on Ethereum run on "*gas*," the fundamental unit of computation in Ethereum. Every transaction, whether to an EOA or to a contract, must have the *gasLimit* and *gasPrice* to compute the fee. This fee is paid to the miners to compensate them for their resource contributions and work they perform. Obviously, miners have the choice of including the transaction and collecting the fee, similar to that of Bitcoin.

Usually, a computational step costs just one *gas*, but some of the compute- or storage-intensive operations cost more. For every byte of transaction data, around five *gas* is required. Take a look at these sample examples: adding two numbers (with EVM opcode ADD) requires approximately three *gas*; multiplying two numbers (with EVM opcode MUL) requires approximately five *gas*; calculating a hash (SHA3) requires around 30 *gas* (compute-intensive, you see). Storage cost is also computed in similar fashion, but quite expensive for good reasons. As per the design, a transaction can include an unlimited amount of data. It costs 68 *gas* per byte of nonzero transaction data. To store a 256-bit word in a "Contract," approximately 20,000 *gas* is required. You could find more opcodes and their corresponding prices in the Ethereum yellow paper. The cost then would be to just multiply the *gas* required with the *gasPrice*. Unlike Bitcoin, Ethereum cost computation is more complex. It takes into account the costs of bandwidth, storage, and computation. Having such a fee computation mechanism prevents the Ethereum network from an attacker who might just want to inject an infinite loop for computation (leading to denial-of-service attacks) or consume more and more space by storing meaningless data.

The total Ether a transaction would cost actually depends on the amount of *gas* consumed by the transaction, multiplied by the price of one unit of *gas* specified in the transaction by the transaction initiator. Miners on the other hand have a strategy for calculating the *gas* Price to charge, which should be the least amount the sender of a transaction must specify so that the transaction does not get rejected by the miner. So, how do you calculate the total cost of a transaction? Not the approximate one, but the actual cost? The total "Ether" cost of a transaction is based on two factors: *gasUsed* and *gasPrice*. Total cost = *gasUsed* * *gasPrice*. The *gasUsed* component is the total *gas* consumed while excuting the EVM opcodes for the instructions, and *gasPrice* is the one specified by the user.

If the total amount of *gas* used by the computational steps (including the transaction, the message, and any submessages that may be triggered) is less than or equal to the *gasLimit*, then the transaction is processed by the miner. However, if the total *gas* exceeds the *gasLimit*, then all changes are reverted (though it is a valid transaction), except that the fee can still be collected by the miner. So, what happens to the excess *gas*? All the unused *gas* after transaction execution is reimbursed to the sender as Ether. Senders do not need to worry about overspending, as they are only charged for the *gas* consumed. This definitely means that it is important as well as safe to send transactions with a *gas* limit well above the estimates. It is also recommended not to pay very high *gas* Price and use the average *gas* price from `https://ethgasstation.info/`.

Let us go through each and every step when a transaction is made in an Ethereun network to build a concrete understanding of the flow:

- Every transaction must define a "*gasLimit*" that it is willing to spend (*gasLimit* is also termed "*startGas*"), and the fee that it is willing to pay per unit of *gas* (*gasPrice*). At the start of execution, Ether worth of *gasLimit* * *gasPrice* is removed from the transaction sender's account. Remember that this is not really the total cost of a transaction (should be a bit more than that in an ideal case). Only after the transaction, its actual cost is concluded (*gasUsed* * *gasPrice*) that's adjusted from this (*gasLimit* * *gasPrice*), which was initially deducted from sender's account and the balance amount is credited back to the sender. In the beginning of a transaction itself, the amount (*gasLimit* * *gasPrice*) is deducted because there could be a possibility that the sender could go bankrupt while the transaction they initiated is midway through.

- All operations during transaction execution, including database reads and writes, messages, and every computational step taken by the E VM such as addition, subtraction, hash, etc. consume a certain quantity of *gas* that is predefined.
- A normal transaction is one that executes successfully without exceeding the *gasLimit* specified. For such transactions, there should be some *gas* remaining, say, "*gas_rem*". After a successful transaction execution, the transaction sender receives a refund of "*gas_rem* * *gasPrice*" and the miner of the block receives a reward of "(*gasLimit* - *gas_rem*) * *gasPrice*".
- If a transaction runs out of *gas* before successful completion, then all executions revert, but the transaction is nevertheless valid. In such situations, the only outcome of the transaction is that the entire amount "*gasLimit* * *gasPrice*" is allocated to the miner.
- In the case of Contract Accounts, when a contract sends a *message* to the other contract for subexecution, it also has the option to set a *gasLimit*. This option is specifically intended for the subexecution arising out of that message, because there is a possibility that the called contract has an infinite loop. If the subexecution runs out of *gas*, then the subexecution is reverted, which protects against such infinite loops or deliberate

attempts of DoS attacks. The *gas* is consumed anyway and allocated to the miner. Also note that when a message is triggered by a contract, only the instructions cost *gas*, but data in a message do not cost any *gas*. This is because the data from the parent contract need not be copied again, and could be just referrenced through a pointer.

The first Ethereum release (Frontier) had a default *gas* Price of 0.05e12 WEI (i.e., smallest denomination of Ether). In the second Ethereum release (Homestead), default *gas* Price was reduced to 0.02e12 WEI. You must be wondering why *gas* and Ether are decoupled from each other and not a single unit of measurement, which would have made it much simpler. Well, it is deliberately designed this way because units of *gas* align well with computation units having a natural cost (e.g., cost per computation), while the price of Ether generally fluctuates as a result of market forces.

We already know that every Ethereum node participating in the network runs the EVM as part of the block verification protocol. This means that all the nodes execute the same set of transactions and contracts (redundantly parallel, but essential for consensus). While this redundency naturally makes it expensive, there is an incentive not to use the blockchain for computation that can be done offchain (Game Theory!).

Typically, 21,000 *gas* is charged for any transaction as a "base fee" to cover the cost of an elliptic curve operation to compute the sender address from the signature, and also for the disk space of storing the transaction. There are ways to estimate *gas* requirements for transactions and contracts. Example: "*estimateGas*" is a Web3 function to estimate *gas* requirement for a given function. Also, to estimate the total cost, *gas price oracle* is a helper function in "geth" client and "web3.eth.getGasPrice" is a Web3 native function to find an approximate *gas* Price. Following is an example code that can be used in "Truffle":

Example code for transaction cost estimation

```
var MyContract = artifacts.require("./MyTest.sol");

// getGasPrice returns the gas price in Wei
MyContract.web3.eth.getGasPrice(function(error, result){
    var gasPrice = Number(result);
    console.log("Current gasPrice is " + gasPrice + " wei");

    // Get the Contract instance
    MyContract.deployed().then(function(instance) {

        // Retrieve gas estimation for the function
        giveAwayDividend()
        return instance.giveAwayDividend.estimateGas(1);

    }).then(function(result) {
        var gas = Number(result);

        console.log("Total gas estimation = " + gas + " units");
        console.log("Total Transaction Cost estimation in Wei =
        " + (gas * gasPrice) + " wei");
        console.log("Total Transaction Cost estimation in
        Ether = " + MyContract.web3.fromWei((gas * gasPrice),
        'ether') + " Ether");
    });
});
```

While writing smart contracts in Solidity, many prefer to use "constant" functions to compute certain things offchain or just make an RPC query to your local blockchain. Since such constant functions do not change the blockchain state, they are in a way free of cost as they do not consume *gas*. If the constant functions are used inside of any transaction, then it is highly likely that *gas* expense would be required.

Let us now learn about the block's *gas* limit. Recollect that Bitcoin had a predefined limit of 1MB block size and Bitcoin cash had a 2MB block size. Miners would accumulate as many transactions as could fit in those blocks. Ethereum, however, has a very different way of limiting the block size. In Ethereum, the block size is controlled by the block *gas* limit. Different transactions have different *gas* limits; so, depending on the block *gas* limit, a certain number of transactions are clubbed together so that total transactions *gas* limit is less than the block *gas* limit. Different miners can have different sets of transactions that they are willing to put in a block. The block *gas* limit is dynamically calculated. The Ethereum protocol allows the miner of a block to adjust the block *gas* limit by a factor of 1/1024 (0.0976%) in either direction. Miners on the Ethereum network use a mining program, such as "ethminer." The ethminer is an Ethereum GPU mining worker, which connects to either *geth* or *Parity* Ethereum client node. Both *geth* and *Parity* have options that miners can change.

Ethereum Smart Contracts

Unlike Bitcoin, which is just a cryptocurrency, Ethereum is so much more–thanks to the smart contracts. We got a glimpse of what a smart contract might be in the previous sections while learning about Contract Accounts. While we will get into the development aspects of smart contracts in the following chapters, we will have a detailed exploration of what they really are in this section.

Let us start with why it is named so? Please be aware that there is nothing "smart" in a smart contract that is out-of-the-box. It is smart when you code smart logic into it, and it is the beauty of Ethereum that enables you to do so. Let us just summarize our learning so far on the Ethereum smart contracts:

- Smart contracts reside inside the Ethereum blockchain.
- They have their own account, hence address and balance.

- They are capable of sending messages and receiving transactions.
- They get activated when they receive a transaction, and can be deactivated as well.
- Like other transactions, an execution fee and storage fee are applicable for them as well.

All the code in Ethereum, including the smart contracts, is compiled to a low level, stack-based bytecode language, referred to as EVM code, that runs on EVM. The popular high-level languages used to write smart contracts are Solidity, Serpent, and LLL, where their respective compilers convert the high-level code into the EVM byte code. We looked at how contracts could be added into the blockchain by any external agent such as EOA. Since computation and storage in Ethereum are very expensive, it is advisible that the logic should be written in as simple and optimized fashion as possible. When a smart contract is deployed to the Ethereum blockchain network, it is possible for anyone to call the functions of the smart contract. The functions usually have security features coded up that prevent unauthorized access; nevertheless, attempts can be made though they won't succeed.

If you try to imagine a smart contract inside of a block in an Ethereum blockchain, it might appear as in Figure 4-9.

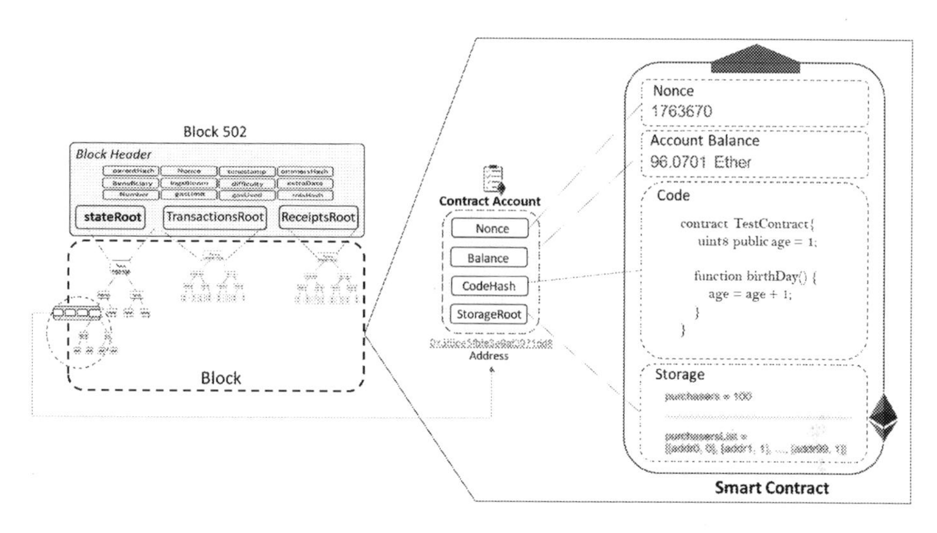

Figure 4-9. *Ethereum smart contract with respect to blocks*

Let us now take an example of a voting application. A smart contract is written that has an address (Contract Account address) and is a part of some block in the blockchain, depending on when it was created. The voters can make transactions to that address (votes). The contract code is written such that it will increment the vote count with every transaction received and terminates itself after some time, publishing the voting result (Ethereum state change). Take a look at Figure 4-10 to have a diagramatic representation for a high-level understanding.

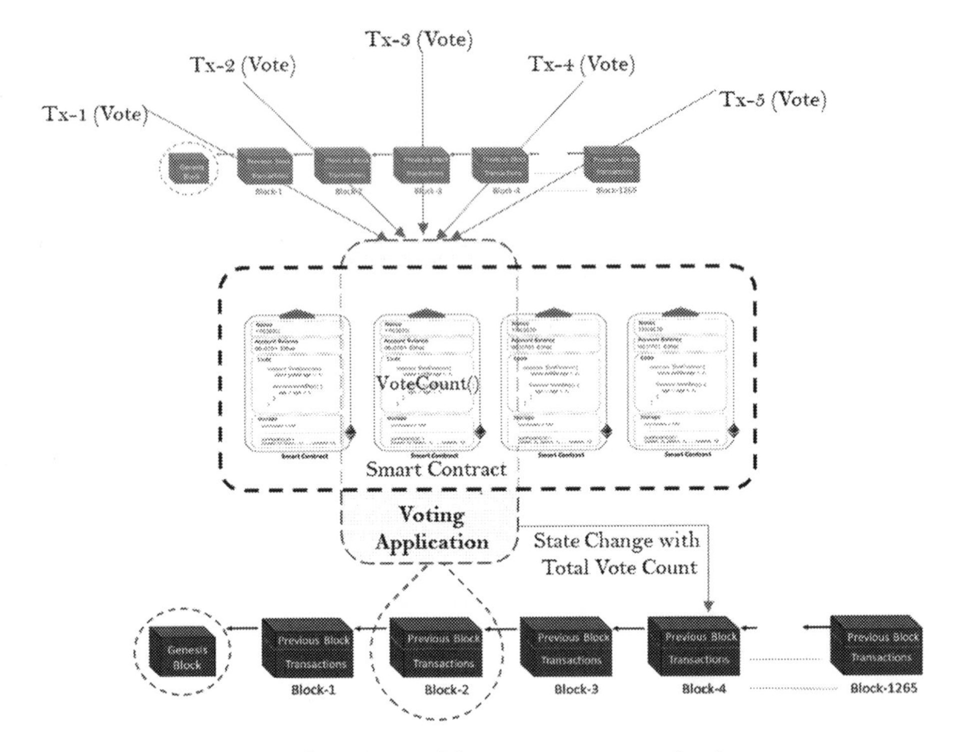

Figure 4-10. *An application with smart contract logic*

Contract Creation

Recollect that we learned about the contract creation transaction, whose only purpose is to create a contract. It is a bit different kind of transaction compared with the other types. So, before the contract creation transaction is fired up to create a Contract Account, it must first initialize the four properties that all types of accounts have:

- The "nonce" should be set to zero initially.
- The "Account Balance" should be set with the value (amount of Ether) transferred by the sender, and the same amount must be deducted from the sender's account.

- The "StorageRoot" should be empty.
- The contract's "codeHash" should be set with the Keccak 256-bit hash of an empty string.

After initializing the account, the account can be created using the *init* code sent with the transaction that does the real work. There could be a whole bunch of actions defined in *init* code, and it's execution can effect several events that are not internal to the execution state, such as:

- The account's 'storage' can be altered.
- Further accounts can be created.
- Further message calls can be triggered.

Ethereum Virtual Machine and Code Execution

Ethereum is a programmable blockchain that allows users to create their own operations of any arbitrary complexity through Turing-complete languages. The EVM is the execution engine of Ethereum that serves as the runtime environment for smart contracts. It is the primary innovation of Ethereum that makes it unique compared with other blockchain systems. It is the EVM on the basis of which the smart contract technology is supposed to get to the next level of innovation, and the game is on. EVM also plays a critical role in transaction execution, changing the state of Ethereum, and achieving consensus. The design goals of EVM are as follows:

- Simplicity: The idea was to make EVM as simple as possible with the low-level constructs. This is why the number of low-level opcodes is kept to a minimum, and so are the data types to the extent that complex logics could still be written conveniently using these constructs. Total 160 instructions, out of which 65 are logically distinct

- Absolute determinism: Ensuring that the execution of instructions with the same set of inputs should produce the same set of outputs (deterministic!) helps maintain the integrity of the EVM without any ambiguity. Determinism along with the concept of "computational step" helps estimate *gas* expense with close approximation.
- Space optimization: In decentralized systems, space saving is a biggest concern. This is why the EVM assembly is kept as compact as possible.
- Tuned for native operations: EVM is tuned for some native operations such as the specific types of arithmatic operations used for cryptography (modular arithmatic), reading blocks or transaction data, interacting with "states," etc. Another such example is: 256-bit (32 bytes) word length to store cryptographic hashes, where EVM operates on the same 256-bits integer.
- Easy security: In a way, *gas* Price helps ensure that the EVM is nonexploitable. If there was no cost, attackers could just keep attacking the system in every possible way. While almost every operation on EVM requires some *gas* cost, it should be easy to come up with good *gas* cost model on EVM

We learned that every participating node in the Ethereum network runs EVM locally, executes all transactions and smart contracts, and saves the final state locally. It is the EVM that writes code (smart contracts) and data to blockchain and executes instructions (opcodes) of transaction code and smart contract code written in a Turing-complete language. That is to say, EVM serves as a runtime environment (RTE) for Ethereum smart contracts and ensures secured execution of the code. Obviously, when the code or transactions are validated through their respective digital signatures, they are executed on EVM. So, only after successful execution of instructions through EVM, the Ethereum state can change.

Unless one connects the EVM with the rest of the network to participate in the P2P network, it can be isolated from the main network. In an isolated and sandboxed environment, EVM could be used to test smart contracts. It facilitates in building better, robust, and production ready-smart contracts.

To build a better understanding of how smart contracts work leveraging the EVM, we should understand how data is organized, stored, and manipulated in any EVM language such as Solidity, Serpent, and ones that might come in future. You might want to consider EVM more like a database engine. Though we will not get deeper into the Solidity programming fundamentals, we will see how it interacts with the EVM in this section. Take a look at Figure 4-11.

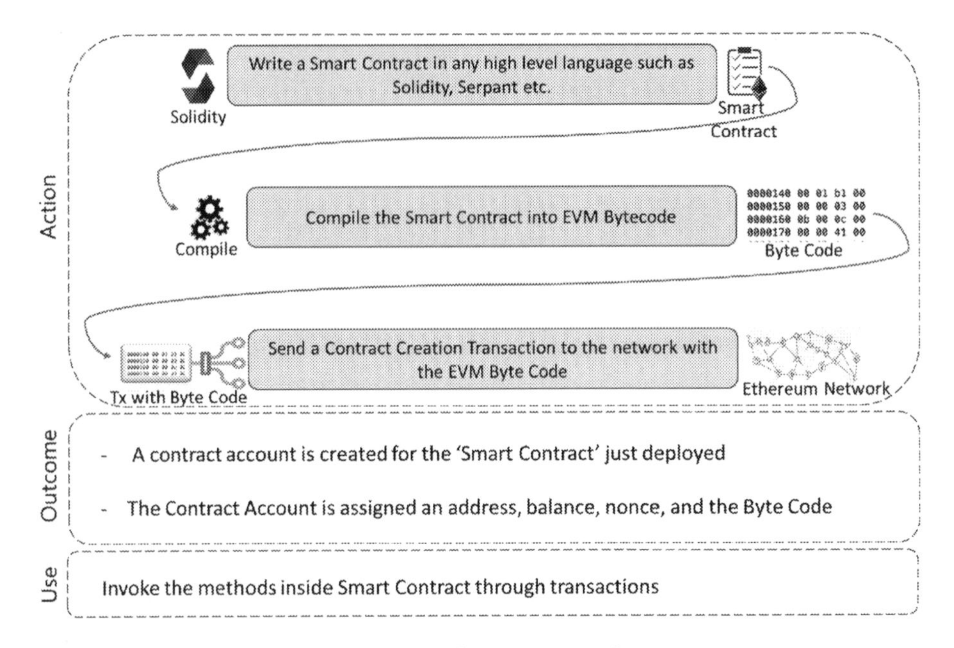

Figure 4-11. *Smart contract deployment and usage*

Let us now understand memory management with EVM. Take a look at the following three strategies that the EVM follows:

- Storage (persistent)
    - Key-value storage mapping (i.e., 256- bit to 256-bit word mapping). This means both keys and values are 256 bits (i.e., 32 bytes).
    - From within a contract, it is not possible to enumarate storage.
    - At any given point in time, the state of the contract can be determined by the contract level variables called "state variables" that are always in "storage," and it cannot be updated at runtime. This means that the structure of the storage is set only once during the contract creation and cannot be altered.

However, their content can be changed with "sendTransaction" calls.

- Read/update of storage is an expensive affair.
- Contracts cannot read, write, or update to any other storage that is not owned by them.
- SSTORE/SLOAD are the frequently used instructions. Example: SSTORE instruction pops the top two items off the stack, considers the first item as the index, and inserts the second item into the contract's storage at that index location.

- Memory (volatile)
    - It is similar to RAM requirement in a general computer system for any code or application execution and used to store temporary values.
    - A contract can use any amount of memory during execution by paying for it, and that memory space is cleaned up after execution completes. The outputs during execution could be pushed to the persistent storage that can be reused in future executions.
    - Memory is actually a byte-array that is contiguous, unlike storage. It is allocated in 256-bit (32 bytes) chunks.
    - Starts with no space and takes on space in the units of 32-byte chunks.
    - Without the "memory" keyword, smart contract languages such as Solidity are expected to declare variables in storage for persistence.

    - Memory cannot be used at the smart contract level; it can only be used in methods.
    - Function arguments are almost always in memory.
    - MSTORE/MLOAD are the frequently used instructions.

- Stack
    - EVM is stack based, hence follows LIFO (Last-in, First-Out), where stack is used to perform computations.
    - Stack entries are also 256-bit words used to mimic 256-bit pseudo registers. They are used to hold local variables of "value" type and to pass parameters to instructions or functions, memory operations, and other algorithmic operations.
    - Allows a maximum of 1024 element and is almost free to use.
    - Most of the stack operations are limited to top of the stacks. The execution is pretty similar to the way Bitcoin script was executed.

When EVM is running and the byte code is injected with a transaction for execution, its full computational state can be defined by the following tuple: [block_state, transaction, message, code, memory, stack, pc, *gas*].

You must be able to make out all these fields now. They have the three kinds of memory we discussed (the block_state field represents the global state and is for storage). The PC field is like a pointer for an instruction in the stack to be executed.

In Ethereum, an Application Binary Interface (ABI) is an abstraction that is not part of the core Ethereum protocol, but is used to access the byte code in a smart contract as standard practice. Though it is possible for anyone to define their own ABI for their contracts and comply with it to get the desired output, it is easier to use Solidity. The purpose of ABI is as follows:

- How and what functions inside smart contracts should be called
- The Binary format in which information should be passed to smart contract functions as inputs
- The Binary format in which you expect the output of function execution after calling that function

With ABI specifications, it is easy (though may not be necessary) for two programs written in two different languages to interact with each other.

Ethereum Ecosystem

We learned the core components to understand how Ethereum really works. There are some inherent limitations to Ethereum such as the following:

- The EVM is slow; it is not advisible to be used for large computations.
- Computation and storage on the blockchain is expensive; it is advisible to use offchain computations and use IPFS/Swarm for storage.
- Scalability is an issue; there are different techniques to address it, but they are subjective to the business case you are dealing with.
- Private blockchains are more likely to flourish.

Now let us take a look at the Ethereum tech stack to understand at a high-level the Ethereum ecosystem.

Swarm

It is not only a distributed storage platform of static files in a P2P fashion, but also a distribution service. Swarm ensures adequate decentralization and redundant storage of Ethereum's blockchain data, DApp code, etc. Unlike WWW, the uploads to Swarm are not centralized to one web server. It is designed to have zero downtime and is DDOS resistant and fault tolerant.

Whisper

It is a communications protocol that allows DApps to communicate with each other. It provides a distributed yet private messaging functionality. It supports singlecast, multicast, and broadcast of messages.

DApp

A DApp usually has two components, a front-end and a back-end component. The back-end code runs on the actual blockchain coded up in smart contracts. The front-end code and user interfaces could be written in any language such as HTML, CSS, and JavaScript, as long as it can make calls to its back end. Also, the front end can be hosted in a decentralized storage like SWARM or IPFS instead of a centralized web server.

User interface components will be cached on some kind of decentralized BitTorrent-like cloud and pulled in by the ĐApp Browser as needed. Like any App store, it is possible to browse the distributed DApps catalog in the browser. The end user can install any DApp of interest in their browser.

Development Components

There are so many development components used to develop decentralized applications on Ethereum and interact with them. Following are a few popular ones, but there are many more such for you to explore. We will just take a look at what they are and dive deeper into these topics in the following chapters.

Web3.js

This is a very important element in developing DApps.

Truffle

Truffle provides the building blocks to create, compile, deploy, and test blockchain applications.

Mist Wallet

We learned in the previous chapters that a wallet is required to interact with blockchain applications and the same applies to Ethereum as well. To store, accept, and send Ether, the users need a wallet. Mist Wallet is a UI-based solution that can be used to connect to the Ethereum blockchain. Using Mist wallet, one can create accounts, design and deploy contracts, transfer Ether across accounts, and view transaction details.

Internally, Mist is dependent on the "geth" client (i.e., GoEthereum Client) to perform all the operations seamlessly.

Summary

In this chapter, we covered the core components of Ethereum blockchain and understood the design considerations. We were able to differentiate Ethereum design with that of the Bitcoin blockchain and understood how

Ethereum blockchain facilitates development of different use cases on a single platform. We took a deep dive into the smart contracts and how the Ethereum Virtual Machine (EVM) executes it in a decentralized fashion.

We will explore more into the development aspect of blockchain in general in Chapter 5 and then build a solid understanding of Ethereum development in Chapter 6.

References

Ethereum White Paper

https://github.com/ethereum/wiki/wiki/White-Paper.

Ethereum Yellow Paper

https://ethereum.github.io/yellowpaper/paper.pdf.

How Ethereum Works

https://medium.com/@preethikasireddy/how-does-ethereum-work-anyway-22d1df506369.

Patricia Tree

https://dl.acm.org/citation.cfm?id=321481.

Merkling in Ethereum

https://blog.ethereum.org/2015/11/15/merkling-in-ethereum/.

***Gas* and Transactions in Ethereum**

http://ethdocs.org/en/latest/contracts-and-transactions/account-types-gas-and-transactions.html.

Technical Introduction to Events and Logs in Ethereum

https://media.consensys.net/technical-introduction-to-events-and-logs-in-ethereum-a074d65dd61e.

Ethereum Internals

https://github.com/comaeio/porosity/wiki/Ethereum-Internals.

CHAPTER 5

Blockchain Application Development

In the previous chapters we went into theoretical details about what blockchain is and how the Bitcoin and Ethereum blockchains work. We also looked at the different cryptographic and mathematical algorithms, theorems, and proofs that go into making the blockchain technology.

In this chapter, we will start with how blockchain applications are different than the conventional applications, and then we will dive into how to build applications on blockchains. We will also look at setting up the necessary infrastructure needed to start developing decentralized applications.

Decentralized Applications

The popularity of blockchain technology is mostly driven by the fact that it can potentially solve various real-world problems because it provides more transparency and security (tamper-proof) than conventional technologies. There are a lot of blockchain use cases identified by several startups and community members aimed at solving these problems.

B. Singhal et al., *Beginning Blockchain*, https://doi.org/10.1007/978-1-4842-3444-0_5

To implement these use cases, we create applications that work on top of blockchains. In general, applications that interact with blockchains are referred to as "decentralized applications" or, in short, just DApps or dApps.

To understand DApps better, let's first revisit what a blockchain is. A blockchain or a distributed ledger is basically a special kind of database where the data is not stored at a centralized server, but it is copied at all the participating nodes in the network. Also, the data on blockchains is cryptographically signed, which proves the identity of the entity who wrote that data on the blockchain. To make use of this database to store and retrieve data, we create applications that are called DApps because these applications do not rely on a centralized database but on a blockchain-based decentralized data store. There is no single point of failure or control for these applications.

Let's take an example of a DApp. Let's take a scenario of supply chain where several vendors and logistics partners are involved in the supply chain process of manufactured goods. To use blockchain technology for this supply chain use case, here's what we would do:

- We would need to set up blockchain nodes at each of these vendors so that they can participate in the consensus process on the data shared.
- We would need an interface so that all the participants and users can store, retrieve, verify, and evaluate data on the blockchain. This interface would be used by the manufacturer to enter the information about the goods manufactured; by the logistics partner to enter information about the transfer of goods; by the warehousing vendor to verify if the goods manufactured and the goods transferred are in sync, etc., etc. This interface would be our supply chain DApp.

Another example of a DApp would be a voting system based on blockchains. Using blockchain for voting, we would be able to make the whole process much more transparent and secure because each vote would be cryptographically signed. We would need to create an application that could get a list of candidates for whom voters could vote, and this application would also provide a simple interface to submit and record the votes.

Blockchain Application Development

Before we jump into code, let's first understand some basic concepts around blockchain application development. Generally, we are used to concepts like objects, classes, functions, etc. when we develop conventional software applications. However, when it comes to blockchain applications, we need to understand a few more concepts like *transactions, accounts and addresses, tokens and wallets, inputs,* and *outputs and balances*. The handshake and request/response mechanism between a decentralized application and a blockchain are driven by these concepts.

First, when developing an application based on blockchain, we need to identify how the application data would map to the blockchain data model. For example, when developing a DApp on the Ethereum blockchain, we need to understand how the application state can be represented in terms of Solidity data structures and how the application's behavior can be expressed in terms of Ethereum smart contracts. As we know that all data on a blockchain is cryptographically signed by private keys of the users, we need to identify which entities in our application would have identities or addresses represented on the blockchain. In conventional applications this is generally not the case, because the data is not always signed. For blockchain application we need to define who would be the signers and what data they would sign. For example, in a voting DApp in which every voter cryptographically signs their vote, this is easy to identify. However, imagine a scenario where we need to migrate an existing conventional distributed systems application, having its data stored across multiple SQL tables and databases, to a DApp based on Ethereum blockchain. In this case we need to identify which entities in which table would have their identities and which entities would be *attached* to other identities.

In the next few sections, we will explore Bitcoin and Ethereum application programming using simple code snippets to send some transactions. The purpose of this exercise is to become familiar with the blockchain APIs and common programming practices. For simplicity, we will be using public test networks for these blockchains and we will write code in JavaScript. The reason for selecting JavaScript is, at the time of this writing, we have stable JavaScript libraries available for both blockchains and it will be easier to understand the similarities and differences in the approaches we take while writing code. The code snippets are explained in detail after every logical step and can be understood even if the reader is not familiar with JavaScript programming.

Libraries and Tools

Recall from Chapter 2, that there are a lot of cryptographic algorithms and mathematics used in blockchain technology. Before we send our transactions to blockchains from an application, we need to prepare them. The transaction preparation includes defining accounts and addresses, adding required parameters and values to the transaction objects, and signing using private keys, among a few other things. When developing applications, it's better to use verified and tested libraries for transaction preparation instead of writing code from scratch. Some of the stable libraries for both Bitcoin and Ethereum are available open source, which can be used to prepare and sign transactions and to send them to the blockchain nodes/network. For the purpose of our code exercises, we will be using the *bitcoinjs* JavaScript library for interacting with the

Bitcoin blockchain and the *web3.js* JavaScript library for interacting with the Ethereum blockchain. Both these libraries are available as node.js packages and can be downloaded and integrated using the *npm* commands.

Important Note The code exercises in this chapter are based on **node.js** applications. This is to make sure that the code we write as part of this exercise has a container in which it can run and interact with the other prepackaged libraries (node modules) mentioned. It is nice to have some knowledge about node.js application development, and the reader is encouraged to follow a *getting started* tutorial on *node.js* and *npm*.

Figure 5-1 shows how a DApp interacts with a blockchain.

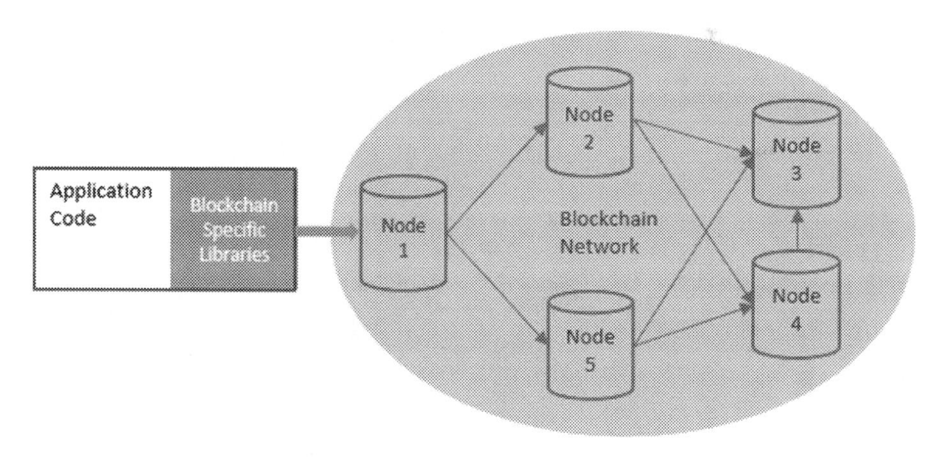

Figure 5-1. *Blockchain application interaction*

Interacting with the Bitcoin Blockchain

In this section we will send a transaction to the Bitcoin public test network from one address to another. Consider this a "*Hello World*" application for the Bitcoin blockchain. As mentioned before, we will be using the *bitcoinjs* JavaScript library for preparing and signing transactions. And for simplicity, instead of hosting a local Bitcoin node, we will use a public Bitcoin test network node hosted by a third-party provider *block-explorer*. Note that you can use any provider for your application and you can also host a local node. All you need to do is to point your application code to connect to your preferred node.

Recall from previous chapters that the Bitcoin blockchain is primarily for enabling peer to peer payments. A Bitcoin transaction is mostly just a transfer of Bitcoins from one address to another. Here's how we do this programmatically.

The following (Figure 5-2) shows how this code interacts with the Bitcoin blockchain. **Note:** The figure is just a rough sketch and does not show the Block Explorer service architecture in detail.

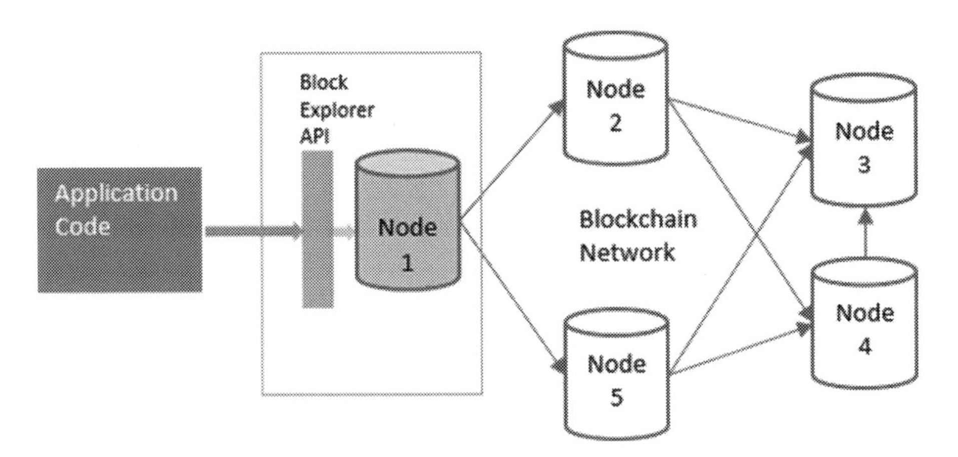

Figure 5-2. *Application interacting with the Bitcoin blockchain using the Block Explorer API*

The following subheadings of this section are steps to follow, in that order, to send a transaction to the Bitcoin test network using JavaScript.

Setup and Initialize the bitcoinjs Library in a *node.js* Application

Before we call the library-specific code for Bitcoin transactions, we will install and initialize the *bitcoinjs* library.

After initializing a node.js applicaion using the **npm init** command, let's create an entry point for our application, *index.js*, and custom JavaScript module to call the bitcoinjs library functions *btc.js*. Import *btc.js* in the *index.js*. Now, we are ready to follow the next steps.

First, let's install the node module for *bitcoinjs*:

```
npm install --save bitcoinjs-lib
```

Then, in our Bitcoin module btc.js, we will initialize the bitcoinjs library using the require keyword:

```
var btc = require('bitcoinjs-lib');
```

Now we can use this btc variable to call library functions on the bitcoinjs library. Also, as part of the initialization process, we are initializing a couple of more variables:

- The network to target : We are using the Bitcoin test network.

  ```
  var network = btc.networks.testnet;
  ```

- The public node API endpoint to get and post transactions : We are using the Block Explorer API for Bitcoin test network. Note that you can replace this API endpoint with your preferred one.

  ```
  var blockExplorerTestnetApiEndpoint =
  'https://testnet.blockexplorer.com/api/';
  ```

At this point, we are all set up to create a Bitcoin transaction using a node.js application.

Create Keypairs for the Sender and Receiver

The first thing that we will need are the keypairs for the sender and the receivers. These are like user accounts identifying the users on the blockchain. So, let's first create two keypairs for Alice and Bob.

```
var getKeys = function () {
    var aliceKeys = btc.ECPair.makeRandom({
        network: network
    });
    var bobKeys = btc.ECPair.makeRandom({
        network: network
    });
    var alicePublic = aliceKeys.getAddress();
    var alicePrivate = aliceKeys.toWIF();
    var bobPublic = bobKeys.getAddress();
    var bobPrivate = bobKeys.toWIF();
    console.log(alicePublic, alicePrivate, bobPublic,
    bobPrivate);
};
```

What we did in the previous code snippet is, we used the **ECPair** class of the bitcoinjs library and called the **makeRandom** method on it to create random keypairs for the test network; note the parameter passed for network type.

Now that we have created a couple of keypairs, let's use them to send Bitcoins from one to the other. In almost all the cryptography examples, Alice and Bob have been the favorite characters, as seen in the preceding keypair variables. However, every time we see a cryptography example, generally Alice is the one who encrypts/signs something and sends to Bob.

For that reason, we feel Bob is under a lot of debt from Alice, so in our case we will help Bob repay some of that debt. We will do this example Bitcoin transaction from Bob to Alice.

Get Test Bitcoins in the Sender's Wallet

We have identified that Bob is going to be acting as the sender in this example Bitcoin transaction. Before he sends any Bitcoins to Alice, he needs to own them. As we know that this example transaction is targeting the Bitcoin test network, there is no real money involved but we still need some test Bitcoins in Bob's wallet. A simple way to get test network Bitcoins is to ask on the Internet. There are a lot of websites on the Internet that host a simple web form to take the Bitcoin testnet addresses and then send test net Bitcoins to those. These services are called Bitcoin testnet faucets, and if you search online for that term you will get a lot of those in the search results. We are not listing or recommending any specific testnet faucet because they are generally not permanent. As soon as a faucet service provider has exhausted their test coins, or they don't want to host the service anymore, they shut it down. But then new ones keep coming up all the time. A list of some of these faucet services is also available on the Bitcoin wiki testnet page.

Another way of getting test net Bitcoins is to host a local Bitcoin node pointing to the test net and mine some. The block mining on the Bitcoin test network is not as difficult as that on the main network. This approach could well be the next level approach when you are building a production Bitcoin application and you need to test it frequently. Instead of asking for test coins every time you want to test your application, you can just mine them yourself.

For the purposes of this simple example, we will just get some Bitcoins from a testnet faucet. In the previous code snippet, the value in the bobPublic variable is Bob's Bitcoin testnet address. When we ran this snippet, it generated "**`msDkUzzd69idLLGCkDFDjVRz44jHcV3pW2`**" as Bob's address. It is also Bob's base 58 encoded public key. We will submit this value in one of the testnet faucet web forms and in return we will receive a transaction ID. If we look up that transaction ID on any of the Bitcoin testnet explorers, we will see that some other address has sent some test Bitcoins to Bob's address we submitted in the form.

Get the Sender's Unspent Outputs

Now that we know that we have some test Bitcoins in Bob's wallet, we can spend them and give them to Alice through a Bitcoin transaction. Let's recall from Chapter 3 how the Bitcoin transactions are made of inputs and outputs. You can spend your unspent outputs by adding them as inputs to the transactions where you want to spend them. To do that, first you need to query the network about the sender's unspent outputs. Here's how we will do that for Bob's Bitcoin testnet address using the block explorer API. To get the unspent outputs, we will send an HTTP request to the UTXO endpoint with Bob's address **"msDkUzzd69idLLGCkDFDjVRz44jHcV3pW2"**.

```
var getOutputs = function () {
    var url = blockExplorerTestnetApiEndpoint + 'addr/' +
    msDkUzzd69idLLGCkDFDjVRz44jHcV3pW2 + '/utxo';
    return new Promise(function (resolve, reject) {
        request.get(url, function (err, res, body) {
            if (err) {
                reject(err);
            }
            resolve(body);
        });
    });
};
```

In the previous code snippet, we have used the node.js request module to send http requests using a node.js application. Feel free to use your favorite http library/module. This snippet is a JavaScript function that returns a promise that resolves into the response body from the API method. Here's how the response looks:

```
[
    {
        address: 'msDkUzzd69idLLGCkDFDjVRz44jHcV3pW2',
        txid: 'db2e5966c5139c6e937203d567403867643482bbd9a6624
        752bbc583ca259958',
        vout: 0,
        scriptPubKey: '76a914806094191cbd4fcd8b4169a70588ad
        c51dc02d6888ac',
        amount: 0.99992,
        satoshis: 99992000,
        height: 1258815,
        confirmations: 1011
    },
    {
      address: 'msDkUzzd69idLLGCkDFDjVRz44jHcV3pW2',
        txid: '5b88d5fc4675bb86b0a3a7fc5a36df9c425c3880a7
        453e3afeb4934e6d1d928e',
        vout: 1,
        scriptPubKey: '76a914806094191cbd4fcd8b4169a70588ad
        c51dc02d6888ac',
        amount: 0.99998,
        satoshis: 99998000,
        height: 1258814,
        confirmations: 1012
    }
]
```

The response body returned by the call is a JSON array with two objects. Each of these objects represents an unspent output for Bob. Each output has txid, which is the transaction ID where this output is listed, the amount associated with output, and the vout, which means the sequence or index number of the output in that transaction. There is some other information in the JSON objects too, but that will not be used in the transaction preparation process.

If we take the first object in the array, it basically says that the Bitcoin testnet address **"msDkUzzd69idLLGCkDFDjVRz44jHcV3pW2"** has `99992000` unspent *satoshis* coming from the transaction `db2e5966c5139c6e937203d567403867643482bbd9a6624752bbc583ca259958` at the index `0`. Similarly, the second object represents `99998000` unspent satoshis coming from the transaction `5b88d5fc4675bb86b0a3a7fc5a36df9c425c3880a7453e3afeb4934e6d1d928e` at the index `1`.

Don't forget that **"msDkUzzd69idLLGCkDFDjVRz44jHcV3pW2"** is Bob's Bitcoin testnet, which we created in step 2 earlier. Now we know that Bob has this many satoshis, which he can spend in a new transaction.

Prepare Bitcoin Transaction

The next step is to prepare a Bitcoin transaction in which Bob can send the test coins to Alice. Preparing the transaction is basically defining its inputs, outputs, and amount.

As we know from the previous step that Bob has two unspent outputs under his Bitcoin testnet address, let's spend the first element of the outputs array. Let's add this as an input to our transaction.

```
var utxo = JSON.parse(body.toString());
var transaction = new btc.TransactionBuilder(network);
transaction.addInput(utxo[0].txid, utxo[0].vout);
```

In the prceding code snippet, first we have parsed the response we received from the previous API call to get Bob's unspent outputs.

Then we have created a transaction builder object for the Bitcoin test network using the bitcoinjs library.

In the last line, we have defined a transaction input. Note that this input is referring to the element at 0 index of the utxo array, which we received in the API call from the previous step. We have passed the transaction ID (txid) and vout from the unspent to the **transaction.addInput** method as input parameters.

Basically, we are defining what we want to spend and where we got it from.

Next, we add the transaction outputs. This is where we say how we want to spend what we added in the input. In the line following, we have added a transaction output by calling the **addOutput** method on the transaction builder object and passed in the target address and the amount. Bob wants to send 99990000 satoshis to Alice. Notice that we have used Alice's Bitcoin testnet address as the function's first parameter.

```
transaction.addOutput(alicePublic, 99990000);
```

While we have used only one input and one output in this example transaction, a transaction can have multiple inputs and outputs. An important thing to note is that the total amount in inputs should not be less than the total amount in outputs. Most of the time, the amount in inputs is slightly more than the amount in outputs, and the difference is the transaction fee offered to the miners to include this transaction when they mine the next block.

In this transaction, we have 2,000 satoshis as the transaction fee, which is the difference between input amount (99992000) and the output amount (99990000). Note that we don't have to create any outputs for the transaction fee; the difference between the input and output total amounts is automatically taken as the transaction fee.

Also, note that we cannot spend partial unspent outputs. If an unspent output has *x* amount of Bitcoins associated with it then we must spend all of the *x* Bitcoins when adding this unspent output as an input in a transaction. So, in case Bob doesn't want to give all the 99,990,000 satoshis associated with his unspent output to Alice, then we need to give it back to Bob by adding another output to the transaction with an amount equal to the difference of total unspent amount and the amount Bob wants to give to Alice.

Sign Transaction Inputs

Now, that we have defined the inputs and outputs in the transaction, we need to sign the inputs using Bob's keys. The following line of code calls the **sign** function on the transaction builder object to cryptographically sign the transaction using Bob's private key, but it takes the whole key pair object as an input parameter.

```
transaction.sign(0, bobKeys);
```

Note that the **transaction.sign** function takes the index of the input and the full key pair as input parameters. In this transaction, because we have only one input, the index we have passed is 0.

At this stage, our transaction is prepared and signed.

Create Transaction Hex

Now we will create a hex string from the transaction object.

```
var transactionHex = transaction.build().toHex();
```

The output of this line of code is the following string, which represents our prepared transaction; this step is needed because the send transaction API accepts the raw transaction as a string.

Broadcast Transaction to the Network

Finally, we use the hex string value we generated in the last step and send it to the block explorer public testnet node using the API,

```
var txPushUrl = blockExplorerTestnetApiEndpoint + 'tx/send';
request.post({
    url: txPushUrl,
        json: {
            rawtx: transactionHex
        }
    }, function (err, res, body) {
        if (err) console.log(err);

        console.log(res);
        console.log(body);
    });
```

If the transaction is accepted by the block explorer public node, we will receive a transaction ID as the response of this API call,

```
{
    txid: "db2e5966c5139c6e937203d567403867643482bbd
    9a6624752bbc583ca259958"
}
```

Now that we have the transaction ID of our transaction, we can look it up on any of the online testnet explorers to see if and when it gets mined and how many confirmations it has.

Putting it all together, here's the complete code for sending a Bitcoin testnet transaction using JavaScript. The input parameters are the Bitcoin testnet keypairs we created in step 1.

```
var createTransaction = function (aliceKeys, bobKeys) {
    getOutputs(bobKeys.getAddress()).then(function (res) {
        var utxo = JSON.parse(res.toString());
        var transaction = new btc.TransactionBuilder(network);
        transaction.addInput(utxo[0].txid, utxo[0].vout);
        transaction.addOutput(alicekeys.getAddress(),
        99990000);
        transaction.sign(0, bobKeys);
        var transactionHex = transaction.build().toHex();
        var txPushUrl = blockExplorerTestnetApiEndpoint +
        'tx/send';
        request.post({
            url: txPushUrl,
            json: {
                rawtx: transactionHex
            }
        }, function (err, res, body) {
            if (err) console.log(err);

            console.log(res);
            console.log(body);
        });
    });
};
```

In this section we learned how we can programmatically send a transaction to the Bitcoin test network. Similarly, we can send transactions to the Bitcoin main network by using the main network as the target in the library functions and in the API endpoints. We also used the query APIs to get unspent outputs of a Bitcoin address. These functions can be used to create a simple Bitcoin wallet application to query and manage Bitcoin addresses and transactions.

Interacting Programmatically with Ethereum—Sending Transactions

The Ethereum blockchain has much more to offer in terms of blockchain application development as compared with the Bitcoin blockchain. The ability to execute logic on the blockchain using smart contracts is the key feature of Ethereum blockchain that allows developers to create decentralized applications. In this section we will learn how to programmatically interact with the Ethereum blockchain using JavaScript. We will look at the main aspects of Ethereum application programming from simple transactions to creating and calling smart contracts.

As we did for interacting with the Bitcoin blockchain in the previous section, we will be using a JavaScript library and test network for interacting with Ethereum as well. We will use the web3 JavaScript library for Ethereum. This library wraps a lot of Ethereum JSON RPC APIs and provides easy to use functions to create Ethereum DApps using JavaScript. At the time of this writing, we are using a version greater than and compatible with version 1.0.0-beta.28 of the web3 JavaScript library.

For the test network, we will be using the *Ropsten* test network for Ethereum blockchain.

For simplicity, we will again use a public-hosted test network node for Ethereum so that we don't have to host a local node while running these code snippets. However, all snippets should work with a locally hosted node as well. We are using the Ethereum APIs provided by the Infura service. Infura is a service that provides public-hosted Ethereum nodes so that developers can easily test their Ethereum apps. There is a small and free registration step needed before we can use the Infura API, so we will go to `https://infura.io` and do a registration. We will get an API key after registration. Using this API key, we can now call the Infura API.

The following (Figure 5-3) shows how this code interacts with the Ethereum blockchain. **Note:** The figure is just a rough sketch and does not show the Infura service architecture in detail.

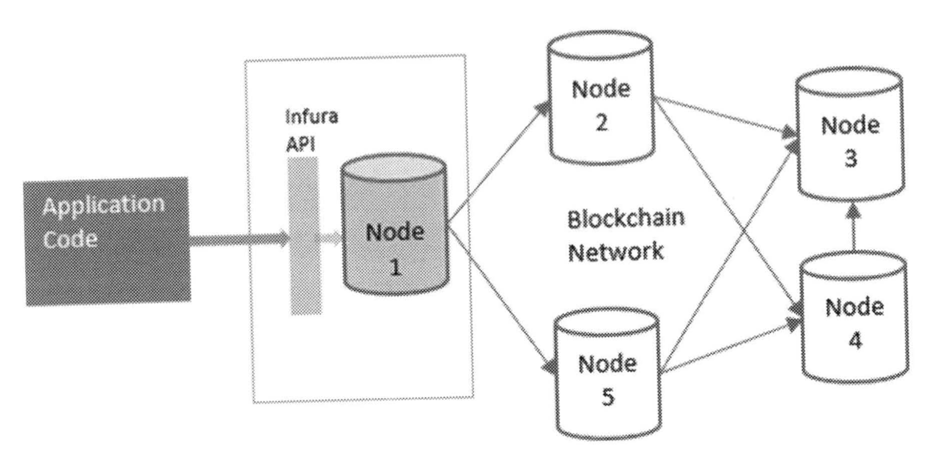

Figure 5-3. *Application interacting with Ethereum blockchain using Infura API service*

The following subsections of this section are steps to follow, in that order, to send a transaction to the Ethereum Ropsten test network using JavaScript.

Set Up Library and Connection

First, we install the web3 library in our node.js application. Note the specific version of library mentioned in the installation command. This is because version 1.0.0 of the library has some more APIs and functions available and they reduce dependency on other external packages.

```
npm install web3@1.0.0-beta.28
```

Then, we initialize the library in our nodejs Ethereum module using the require keyword,

```
var Web3 = require('web3');
```

Now, we have a reference of the web3 library, but we need to instantiate it before we can use it. The following line of code creates a new instance of the Web3 object and it sets the Infura-hosted Ethereum Ropsten test network node as the provider for this Web3 instance.

```
var web3 = new Web3(new Web3.providers.HttpProvider('https://
ropsten.infura.io/<your Infura API key>'));
```

Set Up Ethereum Accounts

Now that we are all set up, let's send a transaction to the Ethereum blockchain. In this transaction, we will send some Ether from one account to another. Recall from Chapter 4 that Ethereum does not use the UTXO model but it uses an account and balances model.

Basically, the Ethereum blockchain manages state and assets in terms of accounts and balances just like banks do. There are no inputs and outputs here. You can simply send Ether from one account to another and Ethereum will make sure that the states are updated for these accounts on all nodes.

To send a transaction to Ethereum that transfers Ether from one account to others, we will first need a couple of Ethereum accounts. Let's start with creating two accounts for Alice and Bob.

The following code snippet calls the account creation function of web3 library and creates two accounts.

```
var createAccounts = function () {
    var aliceKeys = web3.eth.accounts.create();
    console.log(aliceKeys);
    var bobKeys = web3.eth.accounts.create();
    console.log(bobKeys);
};
```

And here's the output that we get in the console window after running the previous snippet.

```
{
    address: '0xAff9d328E8181aE831Bc426347949EB7946A88DA',
    privateKey: '0x9fb71152b32cb90982f95e2b1bf2a5b6b2a5385
    5eacf59d132a2b7f043cfddf5',
    signTransaction: [Function: signTransaction],
    sign: [Function: sign],
    encrypt: [Function: encrypt]
}
{
    address: '0x22013fff98c2909bbFCcdABb411D3715fDB341eA',
    privateKey: '0xc6676b7262dab1a3a28a781c77110b63ab8cd5
    eae2a5a828ba3b1ad28e9f5a9b',
    signTransaction: [Function: signTransaction],
    sign: [Function: sign],
    encrypt: [Function: encrypt]
}
```

As you can see, along with the addresses and private keys, the output for each account creation function call also includes a few functions. For now, we will focus on the address and private key of the returned objects. The address is the Keccak-256 hash of the ECDSA public key of the generated private key. This address and private key combination represents an account on the Ethereum blockchain. You can send Ether to the address and you can spend that Ether using the private key of the corresponding address.

Get Test Ether in Sender's Account

Now, to create an Ethereum transaction which transfers Ether from one account to another, we first need some Ether in one of the accounts. Recall from the Bitcoin programming section that we used testnet faucets to get some test Bitcoins on the address we generated. We will do the same for Ethereum also. Remember that we are targeting the Ropsten test network for Ethereum, so we will search for a Ropsten faucet on the Internet. For this example, we submitted Alice's address that we generated in the previous code snippet to an Ethereum Ropsten test network faucet and we received three ethers on that address.

After receiving Ether on Alice's address, let's check the balance of this address to confirm if we really have the Ether or not. Though we can check the balance of this address using any of the Ethereum explorers online, let's do it using code. The following code snippet calls the getBalance function passing Alice's address as input parameter.

```
var getBalance = function () {
    web3.eth.getBalance('0xAff9d328E8181aE831Bc426347949
    EB7946A88DA').then(console.log);
};
```

And we get the following output as the balance of Alice's address. That's a huge number but that's actually the value of the balance in wei. Wei is the smallest unit of Ether. One Ether equals 10^18 wei. So, the following value equals three Ether, which is what we received from the test network faucet.

```
3000000000000000000
```

Prepare Ethereum Transaction

Now that we have some test Ether with Alice, let's create an Ethereum transaction to send some of this Ether to Bob. Recall that there are no inputs and outputs and UTXO queries to be done in the case of Ethereum because it uses an account and balances-based system. So, all that we need to do in the transaction is to specify the "from" address (the sender's address), the "to" address (the recipient address), and the amount of Ether to be sent, among a few other things.

Also, recall that in the case of a Bitcoin transaction we did not have to specify the transaction fee; however, in the case of an Ethereum transaction we need to specify two related fields. One is *gas* limit and the other is *gas* Price. Recall from Chapter 4 that *gas* is the unit of transaction fee we need to pay to the Ethereum network to get our transactions confirmed and added to blocks. *gas* Price is the amount of Ether (in gwei) we want to pay per unit of *gas*. The maximum fee that we allow to be used for a transaction is the product of *gas* and *gas* Price.

So, for this example transaction, we define a JSON object with the following fields. Here, "from" has Alice's address and "to" has Bob's address, and value is one Ether in wei. The *gas* Price we choose is 20 gwei and the maximum amount of *gas* we want to pay for this transaction is 42,000.

Also, note that we have left the data field empty. We will come back to this later in the smart contract section.

```
{
    from: "0xAff9d328E8181aE831Bc426347949EB7946A88DA",
    gasPrice: "20000000000",
    gas: "42000",
    to: '0x22013fff98c2909bbFCcdABb411D3715fDB341eA',
    value: "1000000000000000000",
    data: ""
}
```

Sign Transaction

Now that we have created a transaction object with the required fields and values, we need to sign it using the private key of the account that is sending the Ether. In this case, the sender is Alice, so we will use Alice's private key to sign the transaction. This is to cryptographically prove that it is actually Alice who is spending the Ether in her account.

```
var signTransaction = function () {
    var tx = {
        from: "0xAff9d328E8181aE831Bc426347949EB7946A88DA",
        gasPrice: "20000000000",
        gas: "42000",
        to: '0x22013fff98c2909bbFCcdABb411D3715fDB341eA',
        value: "1000000000000000000",
        data: ""
    };

    web3.eth.accounts.signTransaction(tx, '0x9fb71152b32cb
    90982f95e2b1bf2a5b6b2a53855eacf59d132a2b7f043cfddf5')
    .then(function(signedTx){
        console.log(signedTx.rawTransaction);
    });
};
```

The preceding code snippet calls the **signTransaction** function with the transaction object we created in the step before and Alice's private key that we got when we generated Alice's account. Following is the output we get when we run the prceding code snippet.

```
{
    messageHash: '0x91b345a38dc728dc06a43c49b92a6ac1e0e6d
    614c432a6dd37d809290a25aa6b',
    v: '0x2a',
    r: '0x14c20901a060834972a539d7b8ad1f23161
    c2144a2b66fbf567e37e963d64537',
    s: '0x3d2a0a818633a11832a5c48708a198af909
    eaf4884a7856c9ac9ed216d9b029c',
```

```
    rawTransaction: '0xf86c018504a817c80082a4109422013fff98c
    2909bbfccdabb411d3715fdb341ea880de0b6b3a76400
    00802aa014c20901a060834972a539d7b8ad1f23161c2144a2b66fbf5
    67e37e963d64537a03d2a0a818633a11832a5c48708a198af909ea
    f4884a7856c9ac9ed216d9b029c'
}
```

In the output of the **signTransaction** function we receive a JSON object with a few properties. The important value for us is the **rawTransaction** value. This is the hex string representation of the signed transaction. This is very similar to how we created a hex string of the Bitcoin transaction in the Bitcoin section.

Send Transaction to the Ethereum Network

The final step is to just send this signed raw transaction to the public-hosted Ethereum test network node, which we have set as the provider of our web3 object.

The following code calls the **sendSignedTransaction** function to send the raw transaction to the Ethereum test network. The input parameter is the value of the **rawTransaction** string that we got in the previous step as part of signing the transaction.

```
web3.eth.sendSignedTransaction(signedTx.rawTransaction).
then(console.log);
```

Notice the use of "then" in the prceding code snippet. This is interesting because the web3 library provides different levels of finality when working with Ethereum transactions, because an Ethereum transaction goes through several states after being submitted. In this function, call of sending a transaction to the network, then, is hit when the transaction receipt is created, and the transaction is complete.

After a few seconds, when the JavaScript promise resolves, the following is what we get as an output.

```
{
    blockHash: '0x26f1e1374d11d4524f692cdf1ce3aa6e085dcc1810
    84642293429eda3954d30e',
    blockNumber: 2514764,
    contractAddress: null,
    cumulativeGasUsed: 125030,
    from: '0xaff9d328e8181ae831bc426347949eb7946a88da',
    gasUsed: 21000,
    logs: [],
    logsBloom: '0x0000000000000000000000000000000000000000000
    000000000000000000000000000000000000000000000000000000000
    000000000000000000000000000000000000000000000000000000000
    000000000000000000000000000000000000000000000000000000000
    000000000000000000000000000000000000000000000000000000000
    000000000000000000000000000000000000000000000000000000000
    000000000000000000000000000000000000000000000000000000000
    000000000000000000000000000000000000000000000000000000000
    000000000000000000000000000000000000000000000000000000000
    0000000000000',
    status: '0x1',
    to: '0x22013fff98c2909bbfccdabb411d3715fdb341ea',
    transactionHash: '0xd3f45394ac038c44c4fe6e0cdb7021fdbd
    672eb1abaa93eb6a1828df5edb6253',
    transactionIndex: 3
}
```

The output has a lot of information, as we can see. The most important part is the **transactionHash**, which is the ID of the transaction on the network. It also gives us the **blockHash**, which is the ID of the block in which this transaction was included. Along with this, we also get information about how much *gas* was used for this transaction, among other details. If the *gas* used is less than the maximum *gas* we specified during transaction creation, the remaining *gas* is sent back to the sender's address.

In this section, we sent a simple transaction to the Ethereum blockchain using JavaScript. But this is just the beginning of Ethereum application programming. In the next section, we will also look at how to create and call smart contracts programmatically.

Interacting Programmatically with Ethereum—Creating a Smart Contract

In this section, we will continue our Ethereum programming exercise, and we will create a simple smart contract on the Ethereum blockchain using the same web3 JavaScript library and the Infura service API.

Because, no computer programming beginners' tutorial is complete without a "Hello World" program, the smart contract we are going to create will be a simple smart contract returning the string "Hello World" when called.

The contract creation process will be a special kind of transaction sent to the Ethereum blockchain, and these types of transactions are called "contract creation transactions." These transactions do not mention a "to" address and the owner of the smart contract is the "from" address mentioned in the transaction.

Prerequisites

In this code exercise to create a smart contract, we will continue with the assumption that the web3 JavaScript library is installed and instantiated in a node.js app and we have registered for the Infura service, just like we did in the previous section.

Following are the steps to create a smart contract on Ethereum using JavaScript.

Program the Smart Contract

Recall from Chapter 4 that the Ethereum smart contracts are written in Solidity programming language. While the web3 JavaScript library will help us deploy our contract on the Ethereum blockchain, we will still have to write and compile our smart contract in Solidity before we send it to the Ethereum network using web3. So, let's first create a sample contract using Solidity.

There are a variety of tools available to code in Solidity. Most of the major IDEs and code editors have Solidity plugins for editing and compiling smart contracts. There is also a web-based Solidity editor called Remix. It's available for free to use at `https://remix.ethereum.org/`. Remix provides a very simple interface to code and compile smart contracts within your browser. In this exercise we will be using Remix to code and test our smart contract and then we will send the same contract to the Ethereum network using the web3 JavaScript library and the Infura API service.

The following code snippet is written in the Solidity programming language and it is a simple smart contract that returns the string "Hello World" from its function Hello. It also has a constructor that sets the value of the message returned.

```
pragma solidity ^0.4.0;
contract HelloWorld {
    string message;
    function HelloWorld(){
        message = "Hello World!";
    }
    function Hello() constant returns (string) {
        return message;
    }
}
```

Let's head to Remix and paste this code in the editor window. The following images (Figures 5-4 and 5-5) show how our sample smart contract looks in the Remix editor and what the output looks like when we clickeded the Create button on the right-side menu, under the Run tab. Also, note that by default, the Remix editor targets a JavaScript VM environment for smart contract compilation and it uses a test account with some ETH balance, for testing purposes. When we click the Create button, this contract is created using the selected account in the JavaScript VM environment.

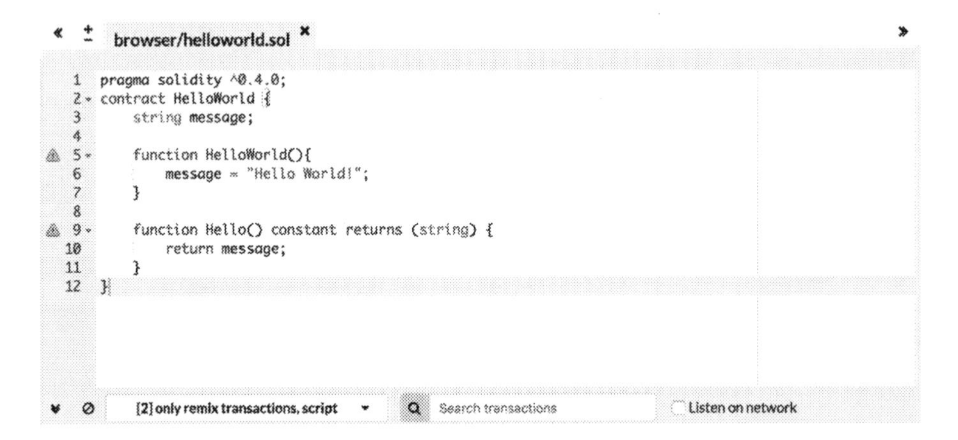

Figure 5-4. *Editing smart contracts in Remix IDE*

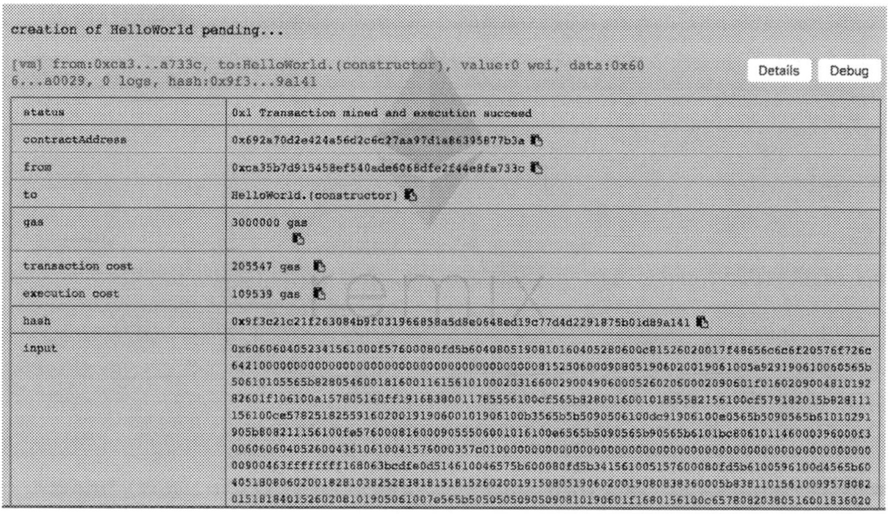

Figure 5-5. *Smart contract creation output in Remix IDE*

Following is the output generated by the create operation, and it shows us that the contract has been created because it has a contract address. The "from" value is the account address that was used to create the contract. It also shows us the hash of the contract creation transaction.

```
status      0x1 Transaction mined and execution succeed
contractAddress     0x692a70d2e424a56d2c6c27aa97d1a86395877b3a
from    0xca35b7d915458ef540ade6068dfe2f44e8fa733c
to      HelloWorld.(constructor)
gas     3000000 gas
transaction cost    205547 gas
execution cost      109539 gas
hash    0x9f3c21c21f263084b9f031966858a5d8e0648ed19c77d4d2291
875b01d89a141
```

```
input  0x6060604052341561000f57600080fd5b6040805190810160405
280600c81526020017f48656c6c6f20576f726c642100000000000000000
000000000000000000000008152506000908051906020019061005a92919
0610060565b50610105565b82805460018160011615610100020316600029
00490600052602060002090601f016020900481019282601f106100a1578
05160ff1916838001178555610cf565b828001600101855582156100cf5
79182015b828111156100ce578251825591602001919060010190610 0b35
65b5b5090506100dc91906100e0565b5090565b61010291905b808211156
100fe5760008160009055506001016100e6565b5090565b90565b6101bc8
06101146000396000f300606060405260043610610041576000357c01000
00000000000000000000000000000000000000000000000000000900463f
fffffff168063bcdfe0d514610046575b600080fd5b341561005157600 08
0fd5b6100596100d4565b6040518080602001828103825283818151815 26
0200191508051906020019080838360005b8381101561009957808201518
18401526020810190506100 7e565b5050505090509081019060 1f1680156
100c65780820380516001836020036101000a03191681526020019150 5b5
09250505060405180910390f35b6100dc61017c565b600080546001816 00
116156101000203166002900480601f016020809104026020016040519 08
101604052809291908181526020018280546001816001161561010002031
66002900480156101725780601f1061014757610100808354040283529 16
0200191610172565b820191906000526020600020905b815481529060010
190602001808311610155578290036 01f168201915b5050505050905090 5
65b6020604051908101604052806000815250905600a165627a7a7230582
0d6796e48540eced3646ea52c632364666e64094479451066317789a712
aef4da0029
 decoded input  {}
 decoded output      -
 logs   []
 value  0 wei
```

At this point, we have a simple *"Hello World"* smart contract ready, and now the next step is to deploy it programmatically to the Ethereum blockchain.

Compile Contract and Get Details

Let's first get some details about our smart contract from Remix, which will be needed to deploy the contract to the Ethereum network using the web3 library. Click on the Compile tab in the right-side menu and then click the Details button. This pops up a new child window with details of the smart contract. What's important for us are the ABI and the BYTECODE sections on the details popup window.

Let's copy the details in the **ABI** section using the *copy value to clipboard* button available next to the ABI header. Following is the value of the ABI data for our smart contract.

```
[
    {
        "constant": true,
        "inputs": [],
        "name": "Hello",
        "outputs": [
            {
                "name": "",
                "type": "string"
            }
        ],
        "payable": false,
        "stateMutability": "view",
        "type": "function"
    },
```

```
    {
        "inputs": [],
        "payable": false,
        "stateMutability": "nonpayable",
        "type": "constructor"
    }
]
```

This is a JSON array and if we closely look at it, we see that it has JSON objects for each function in our contract including its constructor. These JSON objects have details about a function and its input and output. This array describes the smart contract interface.

When we call this smart contract after it is deployed to the network, we will need this information to find out what functions the contract is exposing and what do we need to pass as an input to the function we wish to call.

Now let's get the data in the **BYTECODE** section of the details popup. Following is the data we copied for our contract.


```
{
   "linkReferences": {},
   "object": "606060405234156100 0f57600080fd5b6040805190810
   160405280600c81526020017f48656c6c6f20576f726c64210000000
   0000000000000000000000000000000081525060009080519060200
   19061005a929190610060565b50610105565b8280546001816001161
   5610100020316600290049060005260206000209060 1f01602090048
   1019282601f106100a157805160ff19168380011785556100cf565b8
   2800160010185558215 6100cf579182015b828111156100ce5782518
   2559160200191906001019061 00b3565b5b5090506100dc91906100e
   0565b5090565b61010291905b808211156100fe57600081600090555
   06001016100e6565b5090565b90565b6101bc806101146000396000f
   300606060405260043610610041576000357c0100000000000000000
   000000000000000000000000000000000000000900463ffffffff168
```

```
    063bcdfe0d514610046575b600080fd5b3415610051576000080fd5b6
    100596100d4565b6040518080602001828103825283818151818152602
    001915080519060200190808383600005b838110156100995780820150
    1818401526020810190506100 7e565b5050505090509081019060 1f1
    680156100c65780820380516001836020036101000a03191681526020
    0191505b509250505060405180910390f35b6100dc61017c565b600
    080546001816001161561010002031660029004806 01f01602080910
    402602001604051908101604052809291908181526020018280546 00
    18160011615610100020316600290048015610172578060 1f1061014
    75761010080835404028352916020019161017256 5b820191906000 5
    26020600020905b815481529060010190602001808311610155 57829
    003601f168201915b505050505090509056 5b602060405190810160 4
    052806000815250905600a165627a7a72305820877a5da4f7e05c4ad
    9b45dd10fb6c133a523541ed06db6dd31d59b35d51768a30029",

"opcodes": "PUSH1 0x60 PUSH1 0x40 MSTORE CALLVALUE
ISZERO PUSH2 0xF JUMPI PUSH1 0x0 DUP1 REVERT JUMPDEST
PUSH1 0x40 DUP1 MLOAD SWAP1 DUP2 ADD PUSH1 0x40 MSTORE
DUP1 PUSH1 0xC DUP2 MSTORE PUSH1 0x20 ADD PUSH32
0x48656C6C6F20576F726C6421000000000000000000000000000000
000000 DUP2 MSTORE POP PUSH1 0x0 SWAP1 DUP1 MLOAD SWAP1 PUSH1
0x20 ADD SWAP1 PUSH2 0x5A SWAP3 SWAP2 SWAP1 PUSH2 0x60 JUMP
JUMPDEST POP PUSH2 0x105 JUMP JUMPDEST DUP3 DUP1 SLOAD PUSH1
0x1 DUP2 PUSH1 0x1 AND ISZERO PUSH2 0x100 MUL SUB AND PUSH1
0x2 SWAP1 DIV SWAP1 PUSH1 0x0 MSTORE PUSH1 0x20 PUSH1 0x0
KECCAK256 SWAP1 PUSH1 0x1F ADD PUSH1 0x20 SWAP1 DIV DUP2 ADD
SWAP3 DUP3 PUSH1 0x1F LT PUSH2 0xA1 JUMPI DUP1 MLOAD PUSH1 0xFF
NOT AND DUP4 DUP1 ADD OR DUP6 SSTORE PUSH2 0xCF JUMP JUMPDEST
DUP3 DUP1 ADD PUSH1 0x1 ADD DUP6 SSTORE DUP3 ISZERO PUSH2
0xCF JUMPI SWAP2 DUP3 ADD JUMPDEST DUP3 DUP2 GT ISZERO PUSH2
0xCE JUMPI DUP3 MLOAD DUP3 SSTORE SWAP2 PUSH1 0x20 ADD SWAP2
SWAP1 PUSH1 0x1 ADD SWAP1 PUSH2 0xB3 JUMP JUMPDEST JUMPDEST
```

```
POP SWAP1 POP PUSH2 0xDC SWAP2 SWAP1 PUSH2 0xE0 JUMP JUMPDEST
POP SWAP1 JUMP JUMPDEST PUSH2 0x102 SWAP2 SWAP1 JUMPDEST DUP1
DUP3 GT ISZERO PUSH2 0xFE JUMPI PUSH1 0x0 DUP2 PUSH1 0x0 SWAP1
SSTORE POP PUSH1 0x1 ADD PUSH2 0xE6 JUMP JUMPDEST POP SWAP1
JUMP JUMPDEST SWAP1 JUMP JUMPDEST PUSH2 0x1BC DUP1 PUSH2
0x114 PUSH1 0x0 CODECOPY PUSH1 0x0 RETURN STOP PUSH1 0x60
PUSH1 0x40 MSTORE PUSH1 0x4 CALLDATASIZE LT PUSH2 0x41 JUMPI
PUSH1 0x0 CALLDATALOAD PUSH29 0x1000000000000000000000000000
00000000000000000000000000000 SWAP1 DIV PUSH4 0xFFFFFFFF AND
DUP1 PUSH4 0xBCDFE0D5 EQ PUSH2 0x46 JUMPI JUMPDEST PUSH1 0x0
DUP1 REVERT JUMPDEST CALLVALUE ISZERO PUSH2 0x51 JUMPI PUSH1
0x0 DUP1 REVERT JUMPDEST PUSH2 0x59 PUSH2 0xD4 JUMP JUMPDEST
PUSH1 0x40 MLOAD DUP1 DUP1 PUSH1 0x20 ADD DUP3 DUP2 SUB DUP3
MSTORE DUP4 DUP2 DUP2 MLOAD DUP2 MSTORE PUSH1 0x20 ADD SWAP2
POP DUP1 MLOAD SWAP1 PUSH1 0x20 ADD SWAP1 DUP1 DUP4 DUP4 PUSH1
0x0 JUMPDEST DUP4 DUP2 LT ISZERO PUSH2 0x99 JUMPI DUP1 DUP3
ADD MLOAD DUP2 DUP5 ADD MSTORE PUSH1 0x20 DUP2 ADD SWAP1 POP
PUSH2 0x7E JUMP JUMPDEST POP POP POP POP SWAP1 POP SWAP1 DUP2
ADD SWAP1 PUSH1 0x1F AND DUP1 ISZERO PUSH2 0xC6 JUMPI DUP1
DUP3 SUB DUP1 MLOAD PUSH1 0x1 DUP4 PUSH1 0x20 SUB PUSH2 0x100
EXP SUB NOT AND DUP2 MSTORE PUSH1 0x20 ADD SWAP2 POP JUMPDEST
POP SWAP3 POP POP POP PUSH1 0x40 MLOAD DUP1 SWAP2 SUB SWAP1
RETURN JUMPDEST PUSH2 0xDC PUSH2 0x17C JUMP JUMPDEST PUSH1 0x0
DUP1 SLOAD PUSH1 0x1 DUP2 PUSH1 0x1 AND ISZERO PUSH2 0x100
MUL SUB AND PUSH1 0x2 SWAP1 DIV DUP1 PUSH1 0x1F ADD PUSH1
0x20 DUP1 SWAP2 DIV MUL PUSH1 0x20 ADD PUSH1 0x40 MLOAD SWAP1
DUP2 ADD PUSH1 0x40 MSTORE DUP1 SWAP3 SWAP2 SWAP1 DUP2 DUP2
MSTORE PUSH1 0x20 ADD DUP3 DUP1 SLOAD PUSH1 0x1 DUP2 PUSH1 0x1
AND ISZERO PUSH2 0x100 MUL SUB AND PUSH1 0x2 SWAP1 DIV DUP1
ISZERO PUSH2 0x172 JUMPI DUP1 PUSH1 0x1F LT PUSH2 0x147 JUMPI
```

```
PUSH2 0x100 DUP1 DUP4 SLOAD DIV MUL DUP4 MSTORE SWAP2 PUSH1
0x20 ADD SWAP2 PUSH2 0x172 JUMP JUMPDEST DUP3 ADD SWAP2 SWAP1
PUSH1 0x0 MSTORE PUSH1 0x20 PUSH1 0x0 KECCAK256 SWAP1 JUMPDEST
DUP2 SLOAD DUP2 MSTORE SWAP1 PUSH1 0x1 ADD SWAP1 PUSH1 0x20
ADD DUP1 DUP4 GT PUSH2 0x155 JUMPI DUP3 SWAP1 SUB PUSH1 0x1F
AND DUP3 ADD SWAP2 JUMPDEST POP POP POP POP POP SWAP1 POP
SWAP1 JUMP JUMPDEST PUSH1 0x20 PUSH1 0x40 MLOAD SWAP1 DUP2
ADD PUSH1 0x40 MSTORE DUP1 PUSH1 0x0 DUP2 MSTORE POP SWAP1
JUMP STOP LOG1 PUSH6 0x627A7A723058 KECCAK256 DUP8 PUSH27
0x5DA4F7E05C4AD9B45DD10FB6C133A523541ED0
6DB6DD31D59B35D5 OR PUSH9 0xA30029000000000000 ",
    "sourceMap": "24:199:0:-;;;75:62;;;;;;;;106:24;;;;;;;;;;;;
    ;;;;;;:7;:24;;;;;;;;;;;;:::i;:::-;;24:199;;;;;;;;;;;;;;;;;
    ;;;;;;;;;;;;;;;;;;;;;;;;;;;;;;;;;;;;;;;;;;;;;;;;;;;;;;;;;;
    ;;;;;;;;;;;;;;;;;;;;;:::i;:::-;;;:::o;:::-;;;;;;;;;;;;;;;;
    ;;;;;;;;;;;:::o;:::-;;;;;;;"
}
```

As we can see, the data in the BYTECODE section is a JSON object. This is basically the output of the compilation of the smart contract. Remix compiled our smart contract using the Solidity compiler and as a result we got the solidity byte code. Now closely examine this JSON and look at the "object" property and its value. This is a hex string that contains the byte code for our smart contract, and we will be sending it in the contract creation transaction in the data field–the same data field that we left blank in the previous example Ethereum transaction between Alice and Bob.

Now we have all the details for our smart contract and we are ready to send it to the Ethereum network.

Deploy Contract to Ethereum Network

Now that we have our smart contract and its details, we need to prepare a transaction that can deploy this contract to the Ethereum blockchain. This transaction preparation will be very similar to the transaction we prepared in the previous section, but it will have a few more properties that are needed to create contracts.

First, we need to create an object of the **web3.eth.Contract** class, which can represent our contract. The following code snippet creates an instance for the said class with a JSON array as an input parameter. This is the same JSON array that we copied from the ABI section of the Remix popup window, showing the details about our smart contract.

```
var helloworldContract = new web3.eth.Contract([{
        "constant": true,
        "inputs": [],
        "name": "Hello",
        "outputs": [{
            "name": "",
            "type": "string"
        }],
        "payable": false,
        "stateMutability": "view",
        "type": "function"
    }, {
        "inputs": [],
        "payable": false,
        "stateMutability": "nonpayable",
        "type": "constructor"
    }]);
```

Now we need to send this contract to the Ethereum network using the **Contract.deploy** method of the web3 library. The following code snippet shows how to do this.

```
helloworldContract
.deploy({
        data: '0x6060604052341561000f57600080fd5b604080519081
        0160405280600c81526020017f48656c6c6f20576f726c6421000
        000000000000000000000000000000000000008152506000908051
        906020019061005a929190610060565b506101055565b828054600
        18160011615610100020316600290049060005260206000209060
        1f016020900481019282601f106100a157805160ff19168380011
        785556100cf565b82800160010185558215610cf579182015b82
        8111156100ce5782518255916020019190600101906100b3565b5
        b5090506100dc91906100e0565b5090565b61010291905b808211
        156100fe5760008160009055506001016100e6565b5090565b905
        65b6101bc806101146000396000f30060606060405260043610610
        041576000357c010000000000000000000000000000000000000000
        0000000000000000900463ffffffff168063bcdfe0d514610046
        575b600080fd5b341561005157600080fd5b6100596100d4565b6
        04051808060200182810382528381815181526020019150805190
        602001908083836000 5b83811015610099578082015181840152 6
        0208101905061007e565b50505050905090810190601f16801561
        00c657808203805160018360200361010000a031916815260200 19
        1505b509250505060405180910390f35b6100dc61017c565b6000
        8054600181600116156101000203166002900480601f016020809
        10402602001604051908101604052809291908181526020018280
        54600181600116156101000203166002900480156101725780601
        f10610147576101008083540402835291602001916101725 65b82
        01919060005260206000209 05b81548152906001019060200 1808
```

```
        31161015557829003601f168201915b5050505050905090565b60
        20604051908101604052806000815250905600a165627a7a72305
        820877a5da4f7e05c4ad9b45dd10fb6c133a523541ed06db6dd31
        d59b35d51768a30029'
    })
    .send({
        from: '0xAff9d328E8181aE831Bc426347949EB7946A88DA',
        gas: 4700000,
        gasPrice: '20000000000000'
    },
    function(error, transactionHash){
        console.log(error);
        console.log(transactionHash);
    })
    .then(function(contract){
        console.log(contract);
    });
```

Note that the value of the field data inside the deploy function parameter object is the same value we received in the object field of the BYTECODE details in the previous step. Also notice that the string "0x" is added to this value in the beginning. So, the data passed in the deploy function is '0x' + byte code of the contract.

Inside the send function after the deploy, we have added the "from" address, which will be the owner of the contract and the transaction fee details of *gas* limit and *gas* Price. Finally, when the call is complete, the contract object is returned. This contract object will have the contract details along with the address of the contract, which can be used to call the function on the contract.

Another way of sending the contract to the network would be to wrap the contract inside a transaction and send it directly. The following code snippet creates a transaction object with data as the contract bytecode, signs it using the private key of the address in the "from" field, and then sends it to the Ethereum blockchain.

Note that we have not assigned a "to" address in this transaction object, as the address of the contract is unknown before the contract is deployed.

```
var tx = {
        from: "0x22013fff98c2909bbFCcdABb411D3715fDB341eA",
        gasPrice: "20000000000",
        gas: "4900000",
        data: "0x60606040523415610 00f57600080fd5b604080519081
        0160405280600c81526020017f48656c6c6f20576f726c6421000
        0000000000000000000000000000000000008152506000908051
        906020019061005a929190610060565b50610105565b828054600
        181600116156101000203166002900490600052602060002090 60
        1f016020900481019282601f106100a157805160ff19168380011
        785556100cf565b82800160010185558215610 0cf579182015b82
        8111156100ce578251825591602001919060010190610 0b3565b5
        b5090506100dc91906100e0565b5090565b61010291905b808211
        156100fe5760008160009055506001016100e6565b5090565b905
        65b6101bc806101146000396000f300606060405260043610 6100
        41576000357c01000000000000000000000000000000000000000
        00000000000000000900463ffffffff168063bcdfe0d514610046
        575b600080fd5b341561005157600080fd5b6100596100d4565b6
        040518080602001828103825283818151815260200191508051 90
        602001908083836000 5b83811015610099578082015181840152 6
        020810190506100 7e565b50505050905090810190601f16801561
        00c6578082038051600183602003610100 0a0319168152602001 9
        1505b509250505060405180910390f35b6100dc61017c565b6000
        8054600181600116156101000203166002900480601f016020809
        10402602001604051908101604052809291908181526020018280
        546001816001161561010002031660029004801561017257806 01
```

```
        f106101475761010080835404028352916020019161017 2565b82
        0191906000526020600020905b815481529060010190602001808
        311610155578290036 01f168201915b505050505090509056 5b60
        20604051908101604052806000815250905600a165627a7a72305
        820877a5da4f7e05c4ad9b45dd10fb6c133a523541ed06db6dd31
        d59b35d51768a30029"
    };

    web3.eth.accounts.signTransaction(tx, '0xc6676b7262dab1a3
    a28a781c77110b63ab8cd5eae2a5a828ba3b1ad28e9f5a9b')
    .then(function (signedTx) {
        web3.eth.sendSignedTransaction(signedTx.rawTransaction)
        .then(console.log);
    });
```

When we execute this code snippet, we get the following output, which is the receipt of this transaction.

```
{
    blockHash: '0xaba93b4561fc35e062a1ad72460e0b677603331bbee
    3379ce6c74fa5cf505d82',
    blockNumber: 2539889,
    contractAddress: '0xd5a2d13723A34522EF79bE0f1E7806E86a45
    78E9',
    cumulativeGasUsed: 205547,
    from: '0x22013fff98c2909bbfccdabb411d3715fdb341ea',
    gasUsed: 205547,
    logs: [],
    logsBloom: '0x000000000000000000000000000000000000000000
    00000000000000000000000000000000000000000000000000000
    00000000000000000000000000000000000000000000000000000
    00000000000000000000000000000000000000000000000000000
    00000000000000000000000000000000000000000000000000000
```

```
    0000000000000000000000000000000000000000000000000000000
    0000000000000000000000000000000000000000000000000000000
    0000000000000000000000000000000000000000000000000000000
    0000000000000000000000000000000000000000000000000000000
    0000000000000',
    status: '0x1',
    to: null,
    transactionHash: '0xc333cbc5fc93b52871689aab22c48b910cb19
    2b4875bea69212363030d36565a',
    transactionIndex: 0
}
```

Notice the properties of the transaction receipt object. It has a value assigned to the **contractAddress** property, while the value of the "to" property is null. This means that this was a contract creation transaction that was successfully mined on the network and the contract created as part of this transaction is deployed at the address **`0xd5a2d13723A34522EF79bE0f1E7806E86a4578E9`**.

We have successfully created an Ethereum smart contract programmatically.

Interacting Programmatically with Ethereum—Executing Smart Contract Functions

Now that we have deployed our smart contract to the Ethereum network, we can call its member functions. Following are the steps to call an Ethereum smart contract programmatically.

Get Reference to the Smart Contract

To execute a function of the smart contract, first we need to create an instance of the web3.eth.Contract class with the ABI and address of our deployed contract. The following code snippet shows how to do that.

```
var helloworldContract = new web3.eth.Contract([{
        "constant": true,
        "inputs": [],
        "name": "Hello",
        "outputs": [{
            "name": "",
            "type": "string"
        }],
        "payable": false,
        "stateMutability": "view",
        "type": "function"
    }, {
        "inputs": [],
        "payable": false,
        "stateMutability": "nonpayable",
        "type": "constructor"
    }], '0xd5a2d13723A34522EF79bE0f1E7806E86a4578E9');
```

In the prceding code snippet, we have created an instance of the **web3.eth.Contract** class by passing the ABI of the contract we created in the previous section, and we have also passed the address of the contract that we received after deploying the contract.

This object can now be used to call functions on our contract.

Execute Smart Contract Function

Recall that we have only one public function in our contract. This method is named Hello and it returns the string **"Hello World!"** when executed.

To execute this method, we will call it using the **contract.methods** class in the web3 library. The follwing code snippet shows this.

```
helloworldContract.methods.Hello().send({
        from: '0xF68b93AE6120aF1e2311b30055976d62D7dBf531'
    }).then(console.log);
```

In the prceding code snippet, we have added a value to the "from" address in the send function, and this address will be used to send the transaction that will in turn execute the function Hello on our smart contract.

The full code for calling a smart contract is in the follwing code snippet.

```
var callContract = function () {
    var helloworldContract = new web3.eth.Contract([{
        "constant": true,
        "inputs": [],
        "name": "Hello",
        "outputs": [{
            "name": "",
            "type": "string"
        }],
        "payable": false,
        "stateMutability": "view",
        "type": "function"
    }, {
        "inputs": [],
        "payable": false,
        "stateMutability": "nonpayable",
        "type": "constructor"
    }], '0xd5a2d13723A34522EF79bE0f1E7806E86a4578E9');
```

```
    helloworldContract.methods.Hello().send({
        from: '0xF68b93AE6120aF1e2311b30055976d62D7dBf531'
    }).then(console.log);
};
```

Another way of executing this contract function will be by sending a raw transaction by signing it. It is similar to how we sent a raw Ethereum transaction to send Ether and to create a contract in the previous sections. In this case all we need to do is provide the contract address in the "to" field of the transaction object and the encoded ABI value of the function call in the data field.

The following code snippet first creates a contract object and then gets the encoded ABI value of the smart contract function to be called. It then creates a transaction object based on these values and then signs and sends it to the network. Note that we have used the **encodeABI** function on the contract function to get the data payload value for the transaction. This is the input for the smart contract.

```
var callContract = function () {
    var helloworldContract = new web3.eth.Contract([{
        "constant": true,
        "inputs": [],
        "name": "Hello",
        "outputs": [{
            "name": "",
            "type": "string"
        }],
        "payable": false,
        "stateMutability": "view",
        "type": "function"
    }, {
        "inputs": [],
        "payable": false,
```

```
        "stateMutability": "nonpayable",
        "type": "constructor"
    }], '0xd5a2d13723A34522EF79bE0f1E7806E86a4578E9');

    var payload = helloworldContract.methods.Hello().
    encodeABI();

    var tx = {
        from: "0xF68b93AE6120aF1e2311b30055976d62D7dBf531",
        gasPrice: "20000000000",
        gas: "4700000",
        data: payload
    };

    web3.eth.accounts.signTransaction(tx, '0xc6676b7262dab1a3
    a28a781c77110b63ab8cd5eae2a5a828ba3b1ad28e9f5a9b')
        .then(function (signedTx) {
            web3.eth.sendSignedTransaction(signedTx.raw
            Transaction)
            .then(console.log);
    });
};
```

Important Note When using a public-hosted node for Ethereum, we should use the raw transaction method for creating and executing smart contracts because the web3.eth.Contract submodule of the library uses either an unlocked or default account associated with the provider Ethereum node, but this is not supported by the public nodes (at the time of this writing).

Blockchain Concepts Revisited

In the previous sections we programmatically sent transactions to both Bitcoin and Ethereum blockchains using JavaScript. Here are some of the common concepts that we can now revisit, looking at the process of handcrafting transactions using code.

- **Transactions**: Looking at the code we wrote and the output we got for sending transactions to Ethereum and Bitcoin, we can now say that blockchain transactions are the operations initiated from an account owner, which, if completed successfully, update the state of the blockchain. For example, in our transactions between Alice and Bob, we saw that the ownership of a certain amount of Bitcoins and Ether changed from Alice to Bob and vice versa, and this change of ownership was recorded in the blockchain, hence bringing it into a new state. In the case of Ethereum, transactions go further into contract creation and execution and these transactions also update the state of the blockchain. We created a transaction that in turn deployed a smart contract on the Ethereum blockchain. The state of the blockchain was updated because now we have a new contract account created in the blockchain.
- **Inputs, Outputs, Accounts and Balances**: We also saw how Bitcoin and Ethereum are different from each other in terms of managing the state. While Bitcoin uses the UTXO model, Ethereum uses the accounts and balances model. However, the underlying idea is both the blockchains record the ownership of assets, and transactions are used to change ownership of these assets.

- **Transaction Fee**: For every transaction we do on public blockchain networks, we must pay a transaction fee for our transactions to be confirmed by the miners. In Bitcoin this is automatically calculated, while in Ethereum we should mention the maximum fee we are willing to pay in terms of *gas* Price and *gas* limit.
- **Signing**: In both cases, we also saw that after creating a transaction object with the required values, we signed it using the sender's public key. Cryptographic signing is a way of proving ownership of the assets. If the signature is incorrect, then the transaction becomes invalid.
- **Transaction broadcasting**: After creating and signing the transactions, we sent them to the blockchain nodes. While we sent our example transactions to publicly hosted Bitcoin and Ethereum test network nodes, we are free to send our transactions to multiple nodes if we don't trust all of them to process our transactions. This is called transaction broadcasting.

To summarize, when interacting with blockchains, if we intend to update the state of the blockchain, we submit signed transactions; and to get these transactions confirmed, we need to pay some fee to the network.

Public vs. Private Blockchains

Based on access control, blockchains can be classified as public and private. Public blockchains are also called *permissionless* blockchain and private blockchains are also called *permissioned* blockchains. The primary difference between the two is access control. Public or permissionless blockchains do not restrict addition of new nodes to the network and anyone can join the network. Private blockchains have a limited number of nodes in the network and not everyone can join the network. Examples of public blockchains are Bitcoin and Ethereum main nets. An example of a private blockchain can be a network of a few Ethereum nodes connected to each other but not connected to the main net. These nodes would be collectively called a private blockchain.

Private blockchains are generally used by enterprises to exchange data among themselves and their partners and/or among their suborganizations.

When we develop applications for blockchains, the type of blockchain, public or private, makes a difference because the rules of interaction with the blockchain may or may not be the same. This is called blockchain governance. The public blockchains have a predefined set of rules and the private ones can have a different set of rules per blockchain. A private blockchain for a supply chain may have different governance rules, while a private blockchain for protocol governance may have different rules. For example, the token, *gas* Price, transaction fee, endpoints, etc. may or may not be the same in the aforementioned private Ethereum ledger and the Ethereum main net. This can impact our applications too.

In our code samples, we primarily focused on the public test networks of Bitcoin and Ethereum. While the basic concepts of interacting with private deployments of these blockchains will still be the same, there will be differences in how we configure our code to point to the private networks.

Decentralized Application Architecture

In general, the decentralized applications are meant to directly interact with the blockchain nodes without the need for any centralized components coming into picture. However, in practical scenarios, with legacy systems integrations and limited functionality and scaling of the current blockchain networks, sometimes we must make choices between full decentralization and scalability while designing our DApps.

Public Nodes vs. Self-Hosted Nodes

Blockchains are decentralized networks of nodes. All nodes have the same copy of data and they agree on the state of data always. When we develop applications for blockchains, we can make our application talk to any of the nodes of the target network. There can be mainly two set-ups for this:

- **Application and node both run locally:** The application and the node both run on the local machine. This means we will need our application users to run a local blockchain node and point the application to connect with it. This model would be a purely decentralized model of running an application. An example of this model is the Ethereum-based Mist browser, which uses a local *geth* node.

  Figure 5-6 shows this setup.

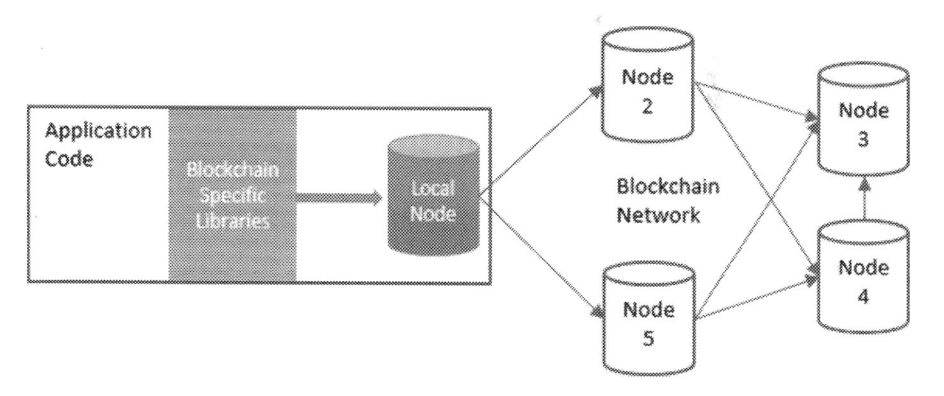

Figure 5-6. *DApp connets to local node*

- **Public node:** The application talks to a public node hosted by a third party. This way our users don't have to host a local node. There are several advantages and disadvantages of this approach. While the users don't have to pay for power and storage for running a local node, they need to trust a third party to broadcast their transactions to the blockchain. The Ethereum browser plugin metamask uses this model and connects with public hosted Ethereum nodes.

Figure 5-7 shows this setup.

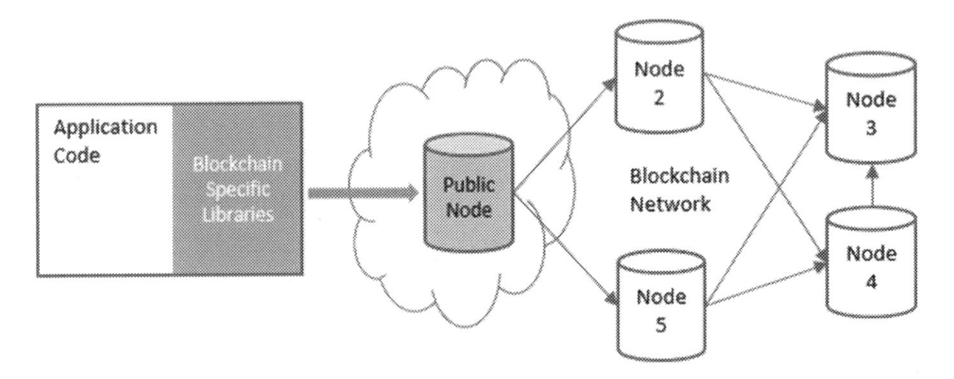

Figure 5-7. *DApp connets to public node*

Decentralized Applications and Servers

Apart from the previously mentioned scenarios, there can be other setups too, depending upon specific use cases and requirements. There are a lot of scenarios when a server is needed between an app and the blockchain. For example: When you need to maintain a cache of the blockchain state for faster queries; when the app needs to send notifications (emails, push, SMS, etc.) to the users based on state updates on the blockchain; and when multiple ledgers are involved, and you need to run a back-end logic to transform data between the ledgers. Imagine the infrastructure being used by some of the big cryptocurrency exchanges where we get all the services like two-factor authentication, notifications, and payment gateways, among other things, and none of these services are available directly in any of the blockchains. In a broader sense, blockchains simply make sure of keeping the data layer tamper resistant and auditable.

Summary

In this chapter we learned about decentralized application development along with some code exercises about interacting programmatically with the Bitcoin and Ethereum blockchains. We also looked at some of the DApp architecture models and how they differ based on the use cases.

In the next chapter we will set up a private Ethereum network and then we will develop a full-fledged DApp interacting with this private network, which will also use smart contracts for business logic.

References

web3.js Documentation

`http://web3js.readthedocs.io/en/1.0/index.html`.

Solidity Documentation

`https://solidity.readthedocs.org/`.

bitcoinjs Source Code Repository

`https://github.com/bitcoinjs/bitcoinjs-lib`.

Infura Documentation

`https://infura.io/docs`.

Block Explorer API Documentation

`https://blockexplorer.com/api-ref`.

Designing the Architecture for your Ethereum Application

`https://blog.zeppelin.solutions/designing-the-architecture-for-your-ethereum-application-9cec086f8317`.

CHAPTER 6

Building an Ethereum DApp

In the previous chapter we learned how to programmatically interact with Bitcoin and Ethereum blockchains using JavaScript. We also touched on how to create and deploy Ethereum smart contracts. In this chapter we will take our blockchain application programming to the next level by learning how to develop and deploy a DApp based on the Ethereum blockchain. As part of creating this DApp, we will be setting up a private Ethereum network and then we will use this network as the underlying blockchain for our DApp. This DApp will have its business logic in an Ethereum smart contract, and this logic will be executed using a web application connecting to private Ethereum network. This way, we intend to cover all aspects of Ethereum application development–from setting up nodes and networks, to creating and deploying a smart contract, to executing smart contract functions using client applications.

The DApp

Before we jump into developing the DApp, we need to define the use case for the DApp. We also need to define the various components that will be part of our DApp. So, let's first do this.

B. Singhal et al., *Beginning Blockchain*, https://doi.org/10.1007/978-1-4842-3444-0_6

The use case for our DApp is a polling application that can allow voters to vote on a poll published in the public domain. Voting using a centralized system is not very reliable, as it exposes a single point of data corruption and failure. So, the goal of our DApp is to enable decentralized polling. This way every voter is in control of their vote and each vote is processed on every node on the blockchain so there is no way to tamper with the vote data. While this can be easily done using the public Ethereum blockchain, to make our exercise interesting we will deploy our polling DApp on a private Ethereum network, and for that we will set up the private network too. Sounds interesting? Let's do this.

The first step will be to set up a private Ethereum network. Then, for hosting the business logic and poll results, we will create a smart contract that will be deployed on this private Ethereum network. To interact with this smart contract, we will create a front-end web application using the web3 library. That's it.

As per the plan just described, our DApp development exercise will have the following steps:

1. Setting up a private Ethereum network
2. Creating a smart contract for polling functionality
3. Deploying the smart contract to the private network
4. Creating a front-end web app to interact with the smart contract

In the following sections, we will be looking, in detail, at each of the steps mentioned.

Note As mentioned, we can also use the public Ethereum network for this DApp development. In addition to that, we can also use several tools like Metamask and Truffle framework to expedite the development of an Ethereum DApp. These tools, along with various others, allow us to manage our code and deployments in a better way. The reader is encouraged to explore these and other tools to try to find the best combination to create a comfortable and productive development environment for their DApp development. This text is primarily focussed on making the reader understand what goes under the hood when creating an Ethereum DApp, hence all tools providing abstractions on top of the DApp development process are kept out of scope.

Setting Up a Private Ethereum Network

To set up a private Ethereum network, we will need one of the many Ethereum clients available. In simple terms, an Ethereum client is an application that implements the Ethereum blockchain protocol. There are many Ethereum clients available on the Internet today; one of the popular ones is go-ethereum, also known as geth. We will be using geth for our private network set-up. For this exercise, we are using a virtual machine running *Ubuntu Linux version 16.04*.

Install go-ethereum (*geth*)

The first step is to install geth on our local machine. To install geth, we will get the geth executable installer from the official source `https://geth.ethereum.org/downloads/`. This download page at the official geth website lists the installer packages for all major platforms (Windows, macOS, Linux).

Download the installer package for your platform and install geth on your local machine. You can also choose to install geth on a remote (cloud-hosted) server/virtual machine if you do not want to install it on your local machine.

Once geth is successfully installed on your local machine, you can check the installation by running the following command in your terminal/command prompt.

```
geth version
```

Depending on your platform OS and the geth version you have installed, this command should give an output similar to the following:

```
Geth
Version: 1.7.3-stable
Git Commit: 4bb3c89d44e372e6a9ab85a8be0c9345265c763a
Architecture: amd64
Protocol Versions: [63 62]
Network Id: 1
Go Version: go1.9
Operating System: linux
GOPATH=
GOROOT=/usr/lib/go-1.9
```

Create *geth* Data Directory

By default, geth will have its working directory but we will create a custom one so that we can track it easily. Simply create a directory and keep the path to this directory handy.

```
mkdir mygeth
```

Create a *geth* Account

The first thing we need is an Ethereum account that can hold Ether. We will need this account to create our smart contracts and transactions later in the DApp development. We can create a new account using the following command.

```
sudo geth account new --datadir <path to the data directory we created in the previous step>
sudo geth account new --datadir /mygeth
```

Note We are using sudo to avoid any permission issues.

When you run this command, the prompt will ask for a passphrase to lock this account. Enter and confirm the passphrase and then your geth account will be created. Make sure to remember the passphrase you entered; it will be needed to unlock the account later to sign transactions. The address of this account will be shown on the screen. For us, the address of the generated account is **baf735f889d603f0ec6b1030c91d9033e60525c3**. The following screenshot (Figure 6-1) shows this process.

```
author@testserver:~$ sudo geth account new --datadir /mygeth
Your new account is locked with a password. Please give a password. Do not forget this p
assword.
Passphrase:
Repeat passphrase:
Address: {baf735f889d603f0ec6b1030c91d9033e60525c3}
```

Figure 6-1. *Ethereum account setup with geth*

Notice that we have passed the data directory as the parameter for the create account command. This is to make sure that the file containing the account details is created inside our data directory so that it is easy to access the account from the context of this directory. If we do not pass the data directory parameter to the geth commands, then it will automatically take the default location of the data directory (which can be different depending on the platform).

Create *genesis.json* Configuration File

After installing geth and creating a new account, the next step is to define the genesis configuration for our private network. As we have seen in the previous chapters, blockchains have a genesis block that acts as the starting point of the blockchain, and all transactions and blocks are validated against the genesis block. For our private network, we will have a custom genesis block and hence a custom genesis configuration. This configuration defines some key values for the blockchain like difficulty level, *gas* limit for blocks, etc.

The genesis configuration for Ethereum has the following format as a JSON object. Each of the keys of this object is a configuration value that drives the network.

```
{
    "config": {
        "chainId": 3792,
        "homesteadBlock": 0,
        "eip155Block": 0,
        "eip158Block": 0
    },
    "difficulty": "2000",
    "gasLimit": "2100000",
    "alloc": {
        "baf735f889d603f0ec6b1030c91d9033e60525c3":
        { "balance": "9000000000000000000" }
    }
}
```

The JSON object is primarily constituted by a config section having values specific to chainId and block numbers related to some of the forks that have taken place. The important parameter to note here is the chainId, which represents the identifier of the blockchain and helps prevent replay attacks. For our private chain, we have opted for a random chainId 3792. You can choose any number here different from the numbers used by main net (1) and the test nets (2, 3, and 4).

The next important parameter is the difficulty. This value defines how difficult it will be to mine a new block. This value is much higher in the Ethereum main network, but for private networks we can choose a relatively smaller value.

Then there is *gasLimit*. This is the total *gas* limit for a block and not a transaction. A higher value generally means more transactions in each block.

Finally, we have the alloc section. Using this configuration, we can prefund Ethereum accounts with the value in wei. As we can see, we have funded the same Ethereum account that we created in the last step, with 9 Ether.

Run the First Node of the Private Network

To run the first node of the private blockchain, let's first copy the JSON from the previous step and save it as a file named **genesis.json**. For simplicity, we are saving this file in the same directory that we are using as the data directory for geth.

First, we need to initialize geth with the *genesis.json*. This initialization is needed to set the custom genesis configuration for our private network.

CD to the directory where we have saved the *genesis.json* file is

```
cd mygeth
```

The following command will initialize geth with the custom configuration we have defined.

```
sudo geth --datadir "/mygeth" init genesis.json
```

geth will confirm the custom genesis configuration set-up with the output in the following screen-shot (Figure 6-2).

```
author@testserver:~/mygeth$ sudo geth --datadir "/mygeth" init genesis.json
INFO [02-11|17:45:32] Allocated cache and file handles        database=/mygeth/geth/chaindata cache=16 handles=16
INFO [02-11|17:45:32] Writing custom genesis block
INFO [02-11|17:45:32] Successfully wrote genesis state        database=chaindata              hash=294e0a…1f4aa7
INFO [02-11|17:45:32] Allocated cache and file handles        database=/mygeth/geth/lightchaindata cache=16 handles=16
INFO [02-11|17:45:32] Writing custom genesis block
INFO [02-11|17:45:32] Successfully wrote genesis state        database=lightchaindata              hash=294e0a…1f4aa7
```

Figure 6-2. *Initialize geth with configuration in genesis.json*

Next, we need to run geth using the following command and the parameters. We will look into each of these parameters in detail.

```
sudo geth --datadir "/mygeth" --networkid 8956 --ipcdisable
--port 30307 --rpc --rpcapi "eth,web3,personal,net,miner,admin,
debug" --rpcport 8507 --mine --minerthreads=1 --etherbase=0xbaf
735f889d603f0ec6b1030c91d9033e60525c3
```

Let's look at each of the parameters that we gave to the geth command.

datadir: This is to specify the data directory just like we did in the previous steps.

networkid: This is the identifier of the network, which differentiates our private blockchain with other Ethereum networks. This is similar to the chainId we defined in the genesis.json file but provides another layer of differentiation among networks. As we can see, we have used another custom number for this value.

ipcdisable: With this parameter we have disabled the interprocess communication port for geth so that while running multiple geth instances (nodes) on the same local machine we should not encounter any conflicting issues.

port: We have selected a custom value for the port to interact with geth.

rpc, --rpcapi, --rpcport: These three parameters define the configuration for the RPC API exposed by geth. We want to enable it; we want eth,web3,perso nal,net,miner,admin,debug geth APIs exposed over RPC; and we want to run it on a custom port 8507.

mine – minerthreads – etherbase: With these three parameters we are instructing geth to start this node as a miner node, limit the miner process threads to only one (so that we do not consume a lot of CPU power), and send the mining rewards to the Ethereum account that we created in the first step.

That's all the configuration we need at this time to run our first geth node for the private network.

When we run this command with all the parameters, geth will give the following output (as in the screenshot shown in Figure 6-3).

```
author@testserver:~/mygeth$ sudo geth --datadir "/mygeth" --networkid 8956 --ipcdisable --port 30307 --rpc --rpcapi "eth,web3,personal,net,miner,a
dmin,debug" --rpcport 8507 --mine --minerthreads=1 --etherbase=0xbaf735f889d603f0ec6b1030c91d9033e60525c3
INFO [02-11|18:00:55] Starting peer-to-peer node                instance=Geth/v1.7.3-stable-4bb3c89d/linux-amd64/go1.9
INFO [02-11|18:00:55] Allocated cache and file handles          database=/mygeth/geth/chaindata cache=128 handles=1024
WARN [02-11|18:00:55] Upgrading database to use lookup entries
INFO [02-11|18:00:55] Initialised chain configuration           config="{ChainID: 3792 Homestead: 0 DAO: <nil> DAOSupport: false EIP150: <nil> EIP1
55: 0 EIP158: 0 Byzantium: <nil> Engine: unknown}"
INFO [02-11|18:00:55] Database deduplication successful         deduped=0
INFO [02-11|18:00:55] Disk storage enabled for ethash caches    dir=/mygeth/geth/ethash count=3
INFO [02-11|18:00:55] Disk storage enabled for ethash DAGs      dir=/home/author/.ethash count=2
INFO [02-11|18:00:55] Initialising Ethereum protocol            versions="[63 62]" network=8956
INFO [02-11|18:00:55] Loaded most recent local header           number=0 hash=294e0a…1f4aa7 td=2000
INFO [02-11|18:00:55] Loaded most recent local full block       number=0 hash=294e0a…1f4aa7 td=2000
INFO [02-11|18:00:55] Loaded most recent local fast block       number=0 hash=294e0a…1f4aa7 td=2000
INFO [02-11|18:00:55] Regenerated local transaction journal     transactions=0 accounts=0
INFO [02-11|18:00:55] Starting P2P networking
INFO [02-11|18:00:57] UDP listener up                           self=enode://e03b50e9b1b2579904f2bbdff7dd0826bd4e4eb2e225c1d1cb1a765195474d7418f3e8
fbfeefd55bd85722973d17626f0e53208c62e38d1099bb583e702b3b48@[::]:30307
INFO [02-11|18:00:57] RLPx listener up                          self=enode://e03b50e9b1b2579904f2bbdff7dd0826bd4e4eb2e225c1d1cb1a765195474d7418f3e8
fbfeefd55bd85722973d17626f0e53208c62e38d1099bb583e702b3b48@[::]:30307
INFO [02-11|18:00:57] HTTP endpoint opened: http://127.0.0.1:8507
INFO [02-11|18:00:57] Transaction pool price threshold updated price=18000000000
INFO [02-11|18:00:57] Starting mining operation
INFO [02-11|18:00:57] Commit new mining work                    number=1 txs=0 uncles=0 elapsed=142µs
INFO [02-11|18:00:59] Generating DAG in progress                epoch=0 percentage=0 elapsed=793.271ms
INFO [02-11|18:01:00] Generating DAG in progress                epoch=0 percentage=1 elapsed=1.592s
```

Figure 6-3. *Geth run first node*

Note the UDP listener up log statement in the output.

```
INFO [02-11|18:00:57] UDP listener up
self=enode://e03b50e9b1b2579904f2bbdff7dd0826bd4e4eb2e
225c1d1cb1a765195474d7418f3e8fbfeefd55bd85722973d1762
6f0e53208c62e38d1099bb583e702b3b48@[::]:30307
```

This contains the address of the node we just started. To connect other nodes to this node we will need this address. Let's keep it noted at some place. The following line has the extracted address from the previous log statement.

```
enode://e03b50e9b1b2579904f2bbdff7dd0826bd4e4eb2e225c1d1cb1a7
65195474d7418f3e8fbfeefd55bd85722973d17626f0e53208c62e38d1099
bb583e702b3b48@[::]:30307
```

Note the **[::]** before the port number we defined in the command. Let's replace this with the local host IP address if we are running the other node on the same machine, or else replace it with the external IP address of the machine. As we are going to run the other network node on the same machine (for development purposes), we will replace it with the localhost IP address. So, the address of the first node will finally be

```
enode://e03b50e9b1b2579904f2bbdff7dd0826bd4e4eb2e225c1d1cb1a
765195474d7418f3e8fbfeefd55bd85722973d17626f0e53208c62e38d10
99bb583e702b3b48@127.0.0.1:30307
```

Run the Second Node of the Network

There is no network with just one node; it should at least have two nodes. So, let's run another geth instance on the same machine, which will interact with the node we just started, and both these nodes together will form our Ethereum private network.

To run another node, first of all we need another directory that can be set as the data directory of the second node. Let us create one.

```
mkdir mygeth2
```

Now, we will initialize this node also with the same genesis.json configuration we created for the first node. Let's create another copy of this genesis.json file and save it in the new directory we created earlier. Let's also CD to this directory. Now, let's initialize the genesis configuration for the second node.

```
sudo geth --datadir "/mygeth2" init genesis.json
```

And, we will get a similar output as we got for the first node. See the screenshot below (Figure 6-4).

```
author@testserver:~/mygeth2$ sudo geth --datadir "/mygeth2" init genesis.json
WARN [02-11|18:19:15] No etherbase set and no accounts found as default
INFO [02-11|18:19:15] Allocated cache and file handles         database=/mygeth2/geth/chaindata cache=16 handles=16
INFO [02-11|18:19:15] Writing custom genesis block
INFO [02-11|18:19:15] Successfully wrote genesis state         database=chaindata               hash=294e0a…1f4aa7
INFO [02-11|18:19:15] Allocated cache and file handles         database=/mygeth2/geth/lightchaindata cache=16 handles=16
INFO [02-11|18:19:15] Writing custom genesis block
INFO [02-11|18:19:15] Successfully wrote genesis state         database=lightchaindata               hash=294e0a…1f4aa7
```

Figure 6-4. *Geth initialize configuration for second node*

Now our second node is also initialized with the genesis configuration. Let's run it.

For running the second node, we will pass a few different parameters to the geth command. This second node will not run as a miner, so we will skip the last three parameters from the command that we gave to the first node. Also, we want to expose the geth console while running this node, so we will add a parameter for that. The command for running the second node will be

```
sudo geth --datadir "/mygeth2" --networkid 8956 --ipcdisable
--port 30308 --rpc --rpcapi "eth,web3,personal,net,miner,admin,
debug" --rpcport 8508 console
```

As we can see, the data directory and ports have been changed for the second node. We also have added the console flag to the command so we can get the geth console for this node.

When we run this command, the second node will also start running and we will see the following output in the terminal (Figure 6-5).

```
author@testserver:~/mygeth2$ sudo geth --datadir "/mygeth2" --networkid 8956 --ipcdisable --port 30308 --rpc --rpcapi "eth,web3,p
ersonal,net,miner,admin,debug" --rpcport 8508 console
WARN [02-11|18:20:27] No etherbase set and no accounts found as default
INFO [02-11|18:20:27] Starting peer-to-peer node              instance=Geth/v1.7.3-stable-4bb3c89d/linux-amd64/go1.9
INFO [02-11|18:20:27] Allocated cache and file handles        database=/mygeth2/geth/chaindata cache=128 handles=1024
WARN [02-11|18:20:27] Upgrading database to use lookup entries
INFO [02-11|18:20:27] Initialised chain configuration         config="{ChainID: 3792 Homestead: 0 DAO: <nil> DAOSupport: false E
IP150: <nil> EIP155: 0 EIP158: 0 Byzantium: <nil> Engine: unknown}"
INFO [02-11|18:20:27] Disk storage enabled for ethash caches  dir=/mygeth2/geth/ethash count=3
INFO [02-11|18:20:27] Disk storage enabled for ethash DAGs    dir=/home/author/.ethash count=2
INFO [02-11|18:20:27] Initialising Ethereum protocol          versions="[63 62]" network=8956
INFO [02-11|18:20:27] Database deduplication successful       deduped=0
INFO [02-11|18:20:27] Loaded most recent local header         number=0 hash=294e0a…1f4aa7 td=2000
INFO [02-11|18:20:27] Loaded most recent local full block     number=0 hash=294e0a…1f4aa7 td=2000
INFO [02-11|18:20:27] Loaded most recent local fast block     number=0 hash=294e0a…1f4aa7 td=2000
INFO [02-11|18:20:27] Regenerated local transaction journal   transactions=0 accounts=0
INFO [02-11|18:20:27] Starting P2P networking
INFO [02-11|18:20:29] UDP listener up                         self=enode://e7ba0ca600eb92e0d665a95edca226dc824230cab7b0bdc403524
32522075d555b38afa9033deea367377f5b5c53ae0ada377c9795f2840b73ffa1c579433e06@[::]:30308
INFO [02-11|18:20:29] RLPx listener up                        self=enode://e7ba0ca600eb92e0d665a95edca226dc824230cab7b0bdc403524
32522075d555b38afa9033deea367377f5b5c53ae0ada377c9795f2840b73ffa1c579433e06@[::]:30308
INFO [02-11|18:20:29] HTTP endpoint opened: http://127.0.0.1:8508
Welcome to the Geth JavaScript console!

instance: Geth/v1.7.3-stable-4bb3c89d/linux-amd64/go1.9
 modules: admin:1.0 debug:1.0 eth:1.0 miner:1.0 net:1.0 personal:1.0 rpc:1.0 txpool:1.0 web3:1.0

>
```

Figure 6-5. *Geth run second node*

At this time, both our geth nodes are running but they do not know about each other. If we run the **admin.peers** command on the geth console of the second node, we will get an empty array as the result (Figure 6-6).

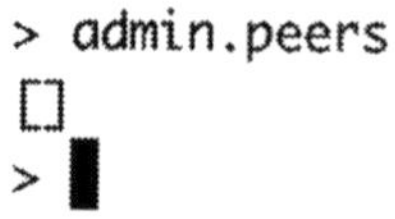

```
> admin.peers
[]
>
```

Figure 6-6. *Geth console–check for peers*

This means that the nodes are not connected to each other. Let's connect the nodes. To do this, we will send the **admin.addPeer()** command on the geth console of the second node with the node address of the first node as the parameter. Remember we noted the address of the first node after running it. Let's run this command in the second node's geth console.

```
admin.addPeer("enode://e03b50e9b1b2579904f2bbdff7dd0826bd4e4e
b2e225c1d1cb1a765195474d7418f3e8fbfeefd55bd85722973d17626f0e5
3208c62e38d1099bb583e702b3b48@127.0.0.1:30307")
```

And as soon as we run this command on the second node, it returns true. Also, after a few seconds it starts synchronization with the first node. The following screen shot (Figure 6-7) shows this output from the console of the second node.

```
> admin.addPeer("enode://e03b50e9b1b2579904f2bbdff7dd0826bd4e4eb2e225c1d1cb1a765195474d741
8f3e8fbfeefd55bd85722973d17626f0e53208c62e38d1099bb583e702b3b48@127.0.0.1:30307")
true
> INFO [02-11|18:34:46] Block synchronisation started
INFO [02-11|18:34:46] Imported new state entries              count=1 elapsed=31µs proces
sed=1 pending=0 retry=0 duplicate=0 unexpected=0
INFO [02-11|18:34:47] Imported new block headers              count=192 elapsed=1.329s nu
mber=192 hash=0939a1…01bc65 ignored=0
INFO [02-11|18:34:47] Imported new block receipts             count=192 elapsed=3.419ms b
ytes=768 number=192 hash=0939a1…01bc65 ignored=0
INFO [02-11|18:34:47] Imported new block headers              count=192 elapsed=306.568ms
 number=384 hash=d4236a…0a25ef ignored=0
```

***Figure 6-7.** Geth console–add peer node*

Both our nodes are now connected and our private Ethereum network is up. To further verify this, we will run the **admin.peers** command again on the second node and this time we will see the JSON array with an object showing the first node as the peer (Figure 6-8).

```
> admin.peers
[{
    caps: ["eth/63"],
    id: "e03b50e9b1b2579904f2bbdff7dd0826bd4e4eb2e225c1d1cb1a765195474d7418f3e8fbfeefd55bd
85722973d17626f0e53208c62e38d1099bb583e702b3b48",
    name: "Geth/v1.7.3-stable-4bb3c89d/linux-amd64/go1.9",
    network: {
      localAddress: "127.0.0.1:53790",
      remoteAddress: "127.0.0.1:30307"
    },
    protocols: {
      eth: {
        difficulty: 120652076,
        head: "0x511dfe69b3e05360ba5deb073e6c34601d6e7266298ccf4c4407aa6683db9285",
        version: 63
      }
    }
}]
```

***Figure 6-8.** Geth console–check for peers (again)*

The following screen shot shows the terminal windows of both the nodes we've set up. On the left is the first node, which is also a miner node, and as we can see it is constantly mining new blocks. The second node is on the right and we can see it is synchronizing with the first node. The screenshot (Figure 6-9) is too small to read because of too much information in it, but it just captures and shows the logs from both Ethereum nodes side by side.

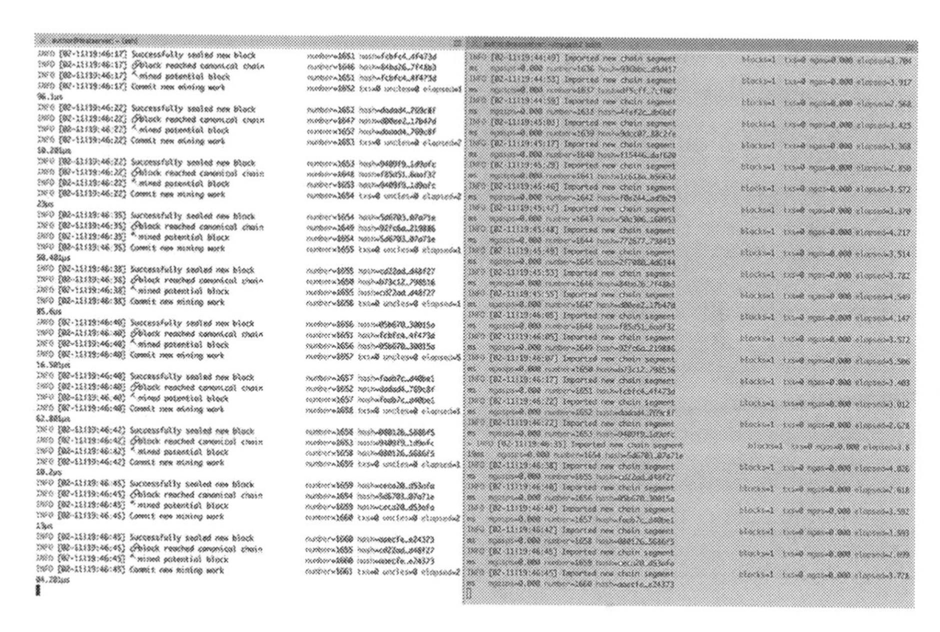

***Figure 6-9.** Geth logs from both Ethereum nodes*

Now that both nodes are peers to each other in the network, we have a working private Ethereum blockchain with two nodes. We also have an Ethereum account that is set as a miner and also is prefunded by some Ether amount. We can now create more accounts and pass Ether among them on this private blockchain.

In this section we learned how to set up a private Ethereum network with two nodes. This can be any number of nodes; we just need to follow the same process for each new node. In case of remote nodes, we should be careful about specifying the right IP addresses of the remote machines and we should also make sure that the required ports are opened if there is a firewall preventing traffic to the machines.

Creating the Smart Contract

Now that we have the private Ethereum network set up and working, we can move on to the next step of creating a smart contract for the polling functionality of our DApp. We will then deploy this contract to our private network. We will follow the same steps of creating and deploying a smart contract as we did in the last chapter.

Let's fire up the Remix online IDE and code our smart contract in Solidity.

The following Solidity code snippet shows the smart contract we have coded for the polling functionality.

```
pragma solidity ^0.4.19;

contract Poll {
    event Voted(
        address _voter,
        uint _value
    );

    mapping(address => uint) public votes;

    string pollSubject = "Should coffee be made tax free? Pass
    1 for yes OR 2 for no in the vote function.";

    function getPoll() constant public returns (string) {
        return pollSubject;
    }

    function vote(uint selection) public {
        Voted(msg.sender, selection);

        require (votes[msg.sender] == 0);
        require (selection > 0 && selection < 3);
        votes[msg.sender] = selection;
    }
}
```

Now, let's analyze this contract source code to understand what we have done here. As we can see, the name of the contract is **Poll**.

The next line of code is

```
event Voted(
    address _voter,
    uint _value
);
```

The preceding code snippet is basically declaring a smart contract event that takes two parameters: one is of the type of Ethereum address and another is of the type of unsigned integer. We have created this event so that we can capture who has voted what in the poll. We will come back to this later.

Next, we have

```
mapping(address => uint) public votes;
```

The preceding line of code declares a mapping of Ethereum addresses and unsigned integers. This is the data store where we will be storing the voters' addresses and their chosen value for the vote.

Then we have the following:

```
string pollSubject = "Should coffee be made tax free? Pass 1
for yes OR 2 for no in the vote function.";
function getPoll() constant public returns (string) {
    return pollSubject;
}
```

The preceding code snippet first declares a string for the polling subject. In this we are asking a question of the voters. And then we have a function that can return the value of this string so that voters can query what the poll is about.

And finally, we have the function that implements the voting functionality.

```
function vote(uint selection) public {
    Voted(msg.sender, selection);

    require (votes[msg.sender] == 0);
    require (selection > 0 && selection < 3);
    votes[msg.sender] = selection;
}
```

Examine closely each line of the preceding snippet.

First, as soon as we enter this function, we are raising the voted event we created with the values of the sender's address (voter) and the value he has chosen.

Next, we are limiting one vote per voter by checking if the value of the vote is zero for the corresponding address in the mapping. The *require* statement is used to check conditions based on user inputs.

And then we are also limiting, by using the require statement, the value of the selection to either 1 or 2. 1 is a yes and 2 is a no. And we have passed these instructions in the pollSubject string so that the voters know what to do.

The screenshot in Figure 6-10 shows the smart contract in Remix,

We compiled this contract code using Remix and we took the ABI and byte code for the contract so that we can deploy it to our private network. We copied the bytecode and ABI from the respective sections in the details popup of the Remix compile tab–exactly how we did this in the previous chapter.

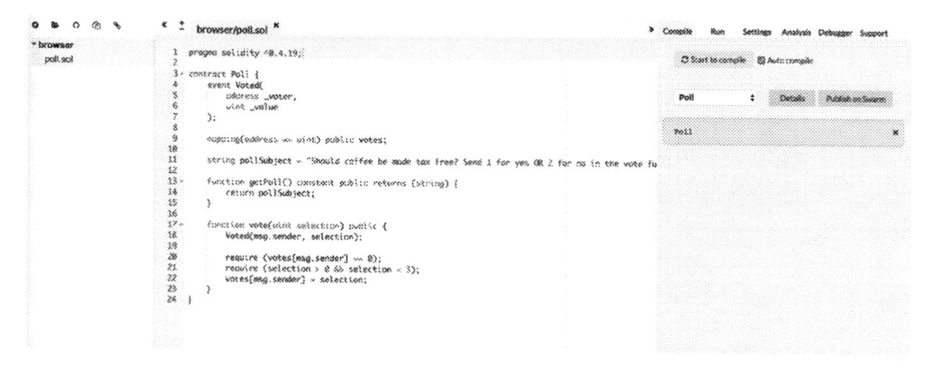

Figure 6-10. *Smart contract editing in Remix online Solidity editor*

The ABI of the contract is

```
[
    {
        "constant": true,
        "inputs": [
            {
                "name": "",
                "type": "address"
            }
        ],
        "name": "votes",
        "outputs": [
            {
                "name": "",
                "type": "uint256"
            }
        ],
        "payable": false,
        "stateMutability": "view",
        "type": "function"
    },
```

```
    {
        "constant": true,
        "inputs": [],
        "name": "getPoll",
        "outputs": [
            {
                "name": "",
                "type": "string"
            }
        ],
        "payable": false,
        "stateMutability": "view",
        "type": "function"
    },
    {
        "anonymous": false,
        "inputs": [
            {
                "indexed": false,
                "name": "_voter",
                "type": "address"
            },
            {
                "indexed": false,
                "name": "_value",
                "type": "uint256"
            }
        ],
        "name": "Voted",
        "type": "event"
    },
```

```
    {
        "constant": false,
        "inputs": [
            {
                "name": "selection",
                "type": "uint256"
            }
        ],
        "name": "vote",
        "outputs": [],
        "payable": false,
        "stateMutability": "nonpayable",
        "type": "function"
    }
]
```

And the byte code for the contract is

```
{
    "linkReferences": {},
    "object": "6060604052608060405190810160405280605081526020
    0017f53686f756c6420636f66666565206265206d616465207461782
    0667265653f2081526020017f53656e64203120666f7220796573204
    f52203220666f72206e6f20696e207468815260200017f6520766f746
    52066756e6374696f6e2e00000000000000000000000000000000815
    25060019080519060200190610009c9291906100ad565b5034156100a
    857600080fd5b610152565b828054600181600116156101000203166
    0029004906000526020600020906001f016020900481019282601f106
    100ee57805160ff1916838001178555561011c565b828001600101855
    5821561011c579182015b828111561011b578251825591602001919
    060010190610100565b5b509050610129919061012d565b5090565b6
    1014f91905b8082111561014b5760008160009055506001016101335
    65b5090565b90565b610373806101616000396000f30060606040526
```

0043610610057576000357c0100000000000000000000000000000000
0000000000000000000000000900463ffffffff1680630121b93f146
1005c57806303c32278146100 7f578063d8bff5a51461010d575b600
080fd5b3415610067576000 80fd5b61007d60048080359060200190 9
190505061015a565b005b341561008a57600080fd5b6100926102735
65b604051808060200182810382528381815181526020019150805 19
0602001908083836000 5b838110156100d2578082015181840152602
0810190506100b7565b50505050905090810190601f16801561 00ff5
7808203805160018360200361010 00a031916815260200191505b509
250505060405180910390f35b341561011857600080fd5b610144600
480803573ffffffffffffffffffffffffffffffffffffffff1690602
0019091905050610 31b565b60405180828152602001915050604051 8
0910390f35b7f4d99b957a2bc29a30ebd96a7be8e68fe50a3c701db2
8a91436490b7d53870ca43382604051808373fffffffffffffffffff
fffffffffffffffffffff1673fffffffffffffffffffffffffffffff
ffffffff168152602001828152602001925050506040518091 0390a
160008060003373ffffffffffffffffffffffffffffffffffffffff1
673ffffffffffffffffffffffffffffffffffffffff1681526020019
0815260200160002054141515610212576000 80fd5b6000811180156
10222575060038110 5b151561022d57600080fd5b806000803373fff
fffffffffffffffffffffffffffffffffffff1673fffffffffffffff
fffffffffffffffffffffffff1681526020019081526020016000208
19055505056 5b61027b610333565b6001805460018160011615610 10
00203166002900480601f0160208091040260200160405190810160 4
05280929190818152602001828054600181600116156101000203166
0029004801561031157806 01f106102e6576101008083540402835 29
1602001916103115 65b820191906000526020600020905b815481529
060010190602001808311610 2f457829003601f168201915b5050505
050905090565b6000602052806000526040600020600091509050548
1565b6020604051908101604052806000815250905600a165627a7a7
2305820ec7d3e1dae8412ec85045a8eafc248e37ae506802cc008ead
300df1ac81aab490029",

```
"opcodes": "PUSH1 0x60 PUSH1 0x40 MSTORE PUSH1 0x80 PUSH1
0x40 MLOAD SWAP1 DUP2 ADD PUSH1 0x40 MSTORE DUP1 PUSH1 0x50
DUP2 MSTORE PUSH1 0x20 ADD PUSH32 0x53686F756C6420636F666
66565206265206D6164652074617820667265653F20 DUP2 MSTORE
PUSH1 0x20 ADD PUSH32 0x53656E64203120666F7220796573204F52
203220666F72206E6F20696E207468 DUP2 MSTORE PUSH1 0x20
ADD PUSH32 0x6520766F74652066756E6374696F6E2E00000000
000000000000000000000000 DUP2 MSTORE POP PUSH1 0x1 SWAP1
DUP1 MLOAD SWAP1 PUSH1 0x20 ADD SWAP1 PUSH2 0x9C SWAP3
SWAP2 SWAP1 PUSH2 0xAD JUMP JUMPDEST POP CALLVALUE ISZERO
PUSH2 0xA8 JUMPI PUSH1 0x0 DUP1 REVERT JUMPDEST PUSH2
0x152 JUMP JUMPDEST DUP3 DUP1 SLOAD PUSH1 0x1 DUP2 PUSH1
0x1 AND ISZERO PUSH2 0x100 MUL SUB AND PUSH1 0x2 SWAP1
DIV SWAP1 PUSH1 0x0 MSTORE PUSH1 0x20 PUSH1 0x0 KECCAK256
SWAP1 PUSH1 0x1F ADD PUSH1 0x20 SWAP1 DIV DUP2 ADD SWAP3
DUP3 PUSH1 0x1F LT PUSH2 0xEE JUMPI DUP1 MLOAD PUSH1 0xFF
NOT AND DUP4 DUP1 ADD OR DUP6 SSTORE PUSH2 0x11C JUMP
JUMPDEST DUP3 DUP1 ADD PUSH1 0x1 ADD DUP6 SSTORE DUP3
ISZERO PUSH2 0x11C JUMPI SWAP2 DUP3 ADD JUMPDEST DUP3 DUP2
GT ISZERO PUSH2 0x11B JUMPI DUP3 MLOAD DUP3 SSTORE SWAP2
PUSH1 0x20 ADD SWAP2 SWAP1 PUSH1 0x1 ADD SWAP1 PUSH2 0x100
JUMP JUMPDEST JUMPDEST POP SWAP1 POP PUSH2 0x129 SWAP2
SWAP1 PUSH2 0x12D JUMP JUMPDEST POP SWAP1 JUMP JUMPDEST
PUSH2 0x14F SWAP2 SWAP1 JUMPDEST DUP1 DUP3 GT ISZERO PUSH2
0x14B JUMPI PUSH1 0x0 DUP2 PUSH1 0x0 SWAP1 SSTORE POP PUSH1
0x1 ADD PUSH2 0x133 JUMP JUMPDEST POP SWAP1 JUMP JUMPDEST
SWAP1 JUMP JUMPDEST PUSH2 0x373 DUP1 PUSH2 0x161 PUSH1
0x0 CODECOPY PUSH1 0x0 RETURN STOP PUSH1 0x60 PUSH1 0x40
MSTORE PUSH1 0x4 CALLDATASIZE LT PUSH2 0x57 JUMPI PUSH1
0x0 CALLDATALOAD PUSH29 0x100000000000000000000000000000
0000000000000000000000000000 SWAP1 DIV PUSH4 0xFFFFFFFF
```

```
AND DUP1 PUSH4 0x121B93F EQ PUSH2 0x5C JUMPI DUP1 PUSH4
0x3C32278 EQ PUSH2 0x7F JUMPI DUP1 PUSH4 0xD8BFF5A5 EQ
PUSH2 0x10D JUMPI JUMPDEST PUSH1 0x0 DUP1 REVERT JUMPDEST
CALLVALUE ISZERO PUSH2 0x67 JUMPI PUSH1 0x0 DUP1 REVERT
JUMPDEST PUSH2 0x7D PUSH1 0x4 DUP1 DUP1 CALLDATALOAD
SWAP1 PUSH1 0x20 ADD SWAP1 SWAP2 SWAP1 POP POP PUSH2 0x15A
JUMP JUMPDEST STOP JUMPDEST CALLVALUE ISZERO PUSH2 0x8A
JUMPI PUSH1 0x0 DUP1 REVERT JUMPDEST PUSH2 0x92 PUSH2
0x273 JUMP JUMPDEST PUSH1 0x40 MLOAD DUP1 DUP1 PUSH1 0x20
ADD DUP3 DUP2 SUB DUP3 MSTORE DUP4 DUP2 DUP2 MLOAD DUP2
MSTORE PUSH1 0x20 ADD SWAP2 POP DUP1 MLOAD SWAP1 PUSH1
0x20 ADD SWAP1 DUP1 DUP4 DUP4 PUSH1 0x0 JUMPDEST DUP4
DUP2 LT ISZERO PUSH2 0xD2 JUMPI DUP1 DUP3 ADD MLOAD DUP2
DUP5 ADD MSTORE PUSH1 0x20 DUP2 ADD SWAP1 POP PUSH2 0xB7
JUMP JUMPDEST POP POP POP POP SWAP1 POP SWAP1 DUP2 ADD
SWAP1 PUSH1 0x1F AND DUP1 ISZERO PUSH2 0xFF JUMPI DUP1
DUP3 SUB DUP1 MLOAD PUSH1 0x1 DUP4 PUSH1 0x20 SUB PUSH2
0x100 EXP SUB NOT AND DUP2 MSTORE PUSH1 0x20 ADD SWAP2 POP
JUMPDEST POP SWAP3 POP POP POP PUSH1 0x40 MLOAD DUP1 SWAP2
SUB SWAP1 RETURN JUMPDEST CALLVALUE ISZERO PUSH2 0x118
JUMPI PUSH1 0x0 DUP1 REVERT JUMPDEST PUSH2 0x144 PUSH1
0x4 DUP1 DUP1 CALLDATALOAD PUSH20 0xFFFFFFFFFFFFFFFFFFFFF
FFFFFFFFFFFFFFFFFFF AND SWAP1 PUSH1 0x20 ADD SWAP1 SWAP2
SWAP1 POP POP PUSH2 0x31B JUMP JUMPDEST PUSH1 0x40 MLOAD
DUP1 DUP3 DUP2 MSTORE PUSH1 0x20 ADD SWAP2 POP POP PUSH1
0x40 MLOAD DUP1 SWAP2 SUB SWAP1 RETURN JUMPDEST PUSH32
0x4D99B957A2BC29A30EBD96A7BE8E68FE50A3C701DB28A91436490B7
D53870CA4 CALLER DUP3 PUSH1 0x40 MLOAD DUP1 DUP4 PUSH20 0xF
FFFFFFFFFFFFFFFFFFFFFFFFFFFFFFFFFFFFFFF AND PUSH20 0xFFFF
FFFFFFFFFFFFFFFFFFFFFFFFFFFFFFFFFFFF AND DUP2 MSTORE PUSH1
0x20 ADD DUP3 DUP2 MSTORE PUSH1 0x20 ADD SWAP3 POP POP POP
```

```
PUSH1 0x40 MLOAD DUP1 SWAP2 SUB SWAP1 LOG1 PUSH1 0x0 DUP1
PUSH1 0x0 CALLER PUSH20 0xFFFFFFFFFFFFFFFFFFFFFFFFFFFFFF
FFFFFFFFFF AND PUSH20 0xFFFFFFFFFFFFFFFFFFFFFFFFFFFFFFFFF
FFFFFFF AND DUP2 MSTORE PUSH1 0x20 ADD SWAP1 DUP2 MSTORE
PUSH1 0x20 ADD PUSH1 0x0 KECCAK256 SLOAD EQ ISZERO ISZERO
PUSH2 0x212 JUMPI PUSH1 0x0 DUP1 REVERT JUMPDEST PUSH1 0x0
DUP2 GT DUP1 ISZERO PUSH2 0x222 JUMPI POP PUSH1 0x3 DUP2
LT JUMPDEST ISZERO ISZERO PUSH2 0x22D JUMPI PUSH1 0x0 DUP1
REVERT JUMPDEST DUP1 PUSH1 0x0 DUP1 CALLER PUSH20 0xFFFFFF
FFFFFFFFFFFFFFFFFFFFFFFFFFFFFFFFFF AND PUSH20 0xFFFFFFFFFF
FFFFFFFFFFFFFFFFFFFFFFFFFFFFFF AND DUP2 MSTORE PUSH1 0x20
ADD SWAP1 DUP2 MSTORE PUSH1 0x20 ADD PUSH1 0x0 KECCAK256
DUP2 SWAP1 SSTORE POP POP JUMP JUMPDEST PUSH2 0x27B PUSH2
0x333 JUMP JUMPDEST PUSH1 0x1 DUP1 SLOAD PUSH1 0x1 DUP2
PUSH1 0x1 AND ISZERO PUSH2 0x100 MUL SUB AND PUSH1 0x2
SWAP1 DIV DUP1 PUSH1 0x1F ADD PUSH1 0x20 DUP1 SWAP2 DIV
MUL PUSH1 0x20 ADD PUSH1 0x40 MLOAD SWAP1 DUP2 ADD PUSH1
0x40 MSTORE DUP1 SWAP3 SWAP2 SWAP1 DUP2 DUP2 MSTORE PUSH1
0x20 ADD DUP3 DUP1 SLOAD PUSH1 0x1 DUP2 PUSH1 0x1 AND
ISZERO PUSH2 0x100 MUL SUB AND PUSH1 0x2 SWAP1 DIV DUP1
ISZERO PUSH2 0x311 JUMPI DUP1 PUSH1 0x1F LT PUSH2 0x2E6
JUMPI PUSH2 0x100 DUP1 DUP4 SLOAD DIV MUL DUP4 MSTORE SWAP2
PUSH1 0x20 ADD SWAP2 PUSH2 0x311 JUMP JUMPDEST DUP3 ADD
SWAP2 SWAP1 PUSH1 0x0 MSTORE PUSH1 0x20 PUSH1 0x0 KECCAK256
SWAP1 JUMPDEST DUP2 SLOAD DUP2 MSTORE SWAP1 PUSH1 0x1
ADD SWAP1 PUSH1 0x20 ADD DUP1 DUP4 GT PUSH2 0x2F4 JUMPI
DUP3 SWAP1 SUB PUSH1 0x1F AND DUP3 ADD SWAP2 JUMPDEST POP
POP POP POP SWAP1 POP SWAP1 JUMP JUMPDEST PUSH1 0x0
PUSH1 0x20 MSTORE DUP1 PUSH1 0x0 MSTORE PUSH1 0x40 PUSH1
0x0 KECCAK256 PUSH1 0x0 SWAP2 POP SWAP1 POP SLOAD DUP2
JUMP JUMPDEST PUSH1 0x20 PUSH1 0x40 MLOAD SWAP1 DUP2 ADD
```

```
    PUSH1 0x40 MSTORE DUP1 PUSH1 0x0 DUP2 MSTORE POP SWAP1
    JUMP STOP LOG1 PUSH6 0x627A7A723058 KECCAK256 0xec PUSH30
    0x3E1DAE8412EC85045A8EAFC248E37AE506802CC008EAD300DF1AC81A
    AB49 STOP 0x29 ",
    "sourceMap": "26:576:0:-;;;167:103;;;;;;;;;;;;;;;;;;;;;;;;;
    ;;;;;;;;;;;;;;;;;:::i;:::-;;26:576;;;;;;;;;;;;;;;;;;;;;;;;;
    ;;;;;;;;;;;;;;;;;;;;;;;;;;;;;;;;;;;;;;;;;;;;;;;;;;;;;;;;;;;
    ;;;;;;;;;;;;;;;;;;;;;;;:::i;:::-;;;:::o;:::-;;;;;;;;;;;;;;;
    ;;;;;;;;;;;;:::o;:::-;;;;;;;"
}
```

With these values our smart contract is ready to be deployed.

Important Note There may be some improvements that can be done to make this contract more secure and performance (*gas*) friendly. The Solidity code should not be taken as reference. Detailed discussion on Solidity best practices is out of scope for this text. For Solidity best practices, we recommend following the official solidity documentation and Solidity-specific texts.

Deploying the Smart Contract

In this section we will deploy the smart contract we developed in the last section to the private Ethereum network we have created. The process of deploying the smart contract is the same as what we did in the previous chapter. The only difference is that this time we are deploying the contract to a private network instead of a public one. In this chapter too, we are using the same web3.js library for Ethereum programming using JavaScript. We recommend the reader to go through the previous chapter if they have not already done that.

Setting up *web3* Library and Connection

First of all, we will install the web3 library in a node.js application. This is exactly how we did it in the last chapter. This node.js application will be used to deploy the smart contract.

```
npm install web3@1.0.0-beta.28
```

After installation, let's first initialize and instantiate the *web3* instance.

```
var Web3 = require('web3');
var web3 = new Web3(new Web3.providers.HttpProvider('ht
tp://127.0.0.1:8507'));
```

Note that this time, our HTTP provider for the web3 instance has changed to a local endpoint instead of a public INFURA endpoint, which we used in the last chapter. This is because we are now connecting to our local private network. Also note that the port we are using is **8507**, which is what we provided in the **--rpcport** parameter when we set up the first node of our private network. This means we are connecting to the first node of the network from our web3 instance.

Deploy the Contract to the Private Network

Now that we have our smart contract and its details, we will prepare a web3 contract object with the details of this contract, and then we will deploy this contract to the Ethereum blockchain by calling the deploy method on the contract object.

We need to create an object of the web3.eth.Contract class that can represent our contract. The following code snippet creates a contract instance with the ABI of our contract as an input to the constructor.

```
var pollingContract = new web3.eth.Contract([
    {
        "constant": true,
        "inputs": [
            {
                "name": "",
                "type": "address"
            }
        ],
        "name": "votes",
        "outputs": [
            {
                "name": "",
                "type": "uint256"
            }
        ],
        "payable": false,
        "stateMutability": "view",
        "type": "function"
    },
    {
        "constant": true,
        "inputs": [],
        "name": "getPoll",
        "outputs": [
            {
                "name": "",
                "type": "string"
            }
        ],
```

```
        "payable": false,
        "stateMutability": "view",
        "type": "function"
    },
    {
        "anonymous": false,
        "inputs": [
            {
                "indexed": false,
                "name": "_voter",
                "type": "address"
            },
            {
                "indexed": false,
                "name": "_value",
                "type": "uint256"
            }
        ],
        "name": "Voted",
        "type": "event"
    },
    {
        "constant": false,
        "inputs": [
            {
                "name": "selection",
                "type": "uint256"
            }
        ],
        "name": "vote",
        "outputs": [],
```

```
        "payable": false,
        "stateMutability": "nonpayable",
        "type": "function"
    }
]);
```

Now we need to deploy this contract to the Ethereum network using the web3 library's deploy method. The following code snippet shows how to do this. In this snippet we have added the byte code in the data field of the object passed to the deploy method.

```
pollingContract
    .deploy({
        data: '0x60606040526080604051908101604052806050815260200
        0017f53686f756c6420636f66666565206265206d6164652074617
        820667265653f2081526020017f53656e64203120666f722079657
        3204f52203220666f72206e6f20696e207468815260200176f65207
        66f74652066756e6374696f6e2e000000000000000000000000000
        000008152506001908051906020019061009c9291906100ad565b5
        034156100a857600080fd5b610152565b82805460018160011615
        6101000203166002900490600052602060002090601f01602090048
        1019282601f106100ee57805160ff1916838001178555561011c565
        b82800160010185555821561011c579182015b828111156101 1b578
        2518255916020019190600101906101 00565b5b509050610129919
        061012d565b5090565b61014f91905b808211156101 4b576000816
        000905550600101610133565b5090565b90565b61037380610161 6
        000396000f300606060405260043610610057576000357c0100000
        000000000000000000000000000000000000000000000000000900
        463ffffffff1680630121b93f1461005c57806303c322781461007
        f578063d8bff5a51461010d575b600080fd5b341561006757600008
        0fd5b61007d6004808035906020019091905050610 15a565b005b3
        41561008a57600080fd5b6100926102 73565b60405180806020018
        2810382528381815181526020019150805190602001908083836 00
```

```
    05b838110156100d25780820151818401526020810190506100b75
    65b50505050905090810190601f1680156100ff578082038051600
    183602003610100a031916815260200191505b5092505050604051
    80910390f35b34156101185760008 0fd5b61014460048080357 3f
    fffffffffffffffffffffffffffffffffffffff169060200190919
    0505061031b565b60405180828152602001915050604051809103 9
    0f35b7f4d99b957a2bc29a30ebd96a7be8e68fe50a3c701db28a91
    436490b7d53870ca43382604051808373fffffffffffffffffffff
    fffffffffffffffffff1673fffffffffffffffffffffffffffffff
    fffffffff168152602001828152602001925050506040518091039
    0a160008060003373fffffffffffffffffffffffffffffffffffff
    fff1673ffffffffffffffffffffffffffffffffffffffff1681526
    020019081526020016000205414151561021257600080fd5b60008
    111801561022257506003811 05b151561022d57600080fd5b80600
    0803373ffffffffffffffffffffffffffffffffffffffff1673fff
    fffffffffffffffffffffffffffffffffffff16815260200190815
    26020016000208190555050565b61027b610333565b60018054600
    181600116156101000203166002900480601f01602080910402602
    001604051908101604052809291908181526020018280546001816
    001161561010002031660029004801561031157806 01f106102e65
    761010080835404028352916020019161031156 5b820191906000 5
    26020600020905b815481529060010190602001808311610 2f4578
    29003601f168201915b505050505090509056 5b6000602052806 00
    0526040600020600091509050548156 5b6020604051908101604 05
    2806000815250905600a165627a7a72305820ec7d3e1dae8412ec8
    5045a8eafc248e37ae506802cc008ead300df1ac81aab490029'
})
.send({
    from: '0xbaf735f889d603f0ec6b1030c91d9033e60525c3',
    gas: 4700000,
    gasPrice: '20000000000000'
},
```

```
    function(error, transactionHash){
        console.log(error);
        console.log(transactionHash);
    })
    .then(function(contract){
        console.log(contract);
    });
```

Note that we have also used the account that we created during the network setup in the "from" field of the send function. As this account was prefunded with nine Ether and it's also added as the etherbase account for the mining rewards, it has enough Ether to deploy a contract.

The full function to deploy the contract will be

```
var deployContract = function () {
    var pollingContract = new web3.eth.Contract([{
            "constant": true,
            "inputs": [{
                "name": "",
                "type": "address"
            }],
            "name": "votes",
            "outputs": [{
                "name": "",
                "type": "uint256"
            }],
            "payable": false,
            "stateMutability": "view",
            "type": "function"
        },
        {
            "constant": true,
            "inputs": [],
```

```
        "name": "getPoll",
        "outputs": [{
            "name": "",
            "type": "string"
        }],
        "payable": false,
        "stateMutability": "view",
        "type": "function"
    },
    {
        "anonymous": false,
        "inputs": [{
                "indexed": false,
                "name": "_voter",
                "type": "address"
            },
            {
                "indexed": false,
                "name": "_value",
                "type": "uint256"
            }
        ],
        "name": "Voted",
        "type": "event"
    },
    {
        "constant": false,
        "inputs": [{
            "name": "selection",
            "type": "uint256"
        }],
```

```
            "name": "vote",
            "outputs": [],
            "payable": false,
            "stateMutability": "nonpayable",
            "type": "function"
        }
    ]);
    pollingContract
        .deploy({
            data: '0x6060604052608060405190810160405280605081 5
            26020017f53686f756c6420636f66666565206265206d61646
            520746178206672656 53f2081526020017f53656e642031206
            66f7220796573204f52203220666f72206e6f20696e2074688
            1526020017f6520766f74652066756e6374696f6e2e0000000
            0000000000000000000000008152506001908051906020019
            061009c9291906100ad565b5034156100a857600080fd5b610
            152565b828054600181600116156101000203166002900490 6
            00052602060002090601f016020900481019282601f106100e
            e57805160ff1916838001178555561011c565b828001600101 8
            55582156101 1c579182015b82811115610 11b5782518255916
            02001919060010190610100565b5b509050610129919061012
            d565b5090565b61014f91905b8082111561014b57600081600
            09055506001016101 33565b5090565b90565b610373806101 6
            16000396000f3006060604052600436106100575760003 57c0
            10000000000000000000000000000000000000000000000
            0000000900463ffffffff1680630121b93f1461005c5780630
            3c32278146100 7f578063d8bff5a51461010d575b600080fd5
            b3415610067576 00080fd5b61007d6004808035906020019 09
            190505061015a565b005b341561008a57600080fd5b6100926
            10273565b6040518080602001828103825283818151815260 2
```

```
    00191508051906020019080838360005b838110156100d2578
    08201518184015260208101905061 00b7565b505050509050 9
    0810190601f1680156100ff5780820380516001836020036 10
    1000a031916815260200191505b5092505050604051809103 9
    0f35b34156101185760008 0fd5b61014460048 0803573fffff
    fffffffffffffffffffffffffffffffffff169060200190919
    0505061031b565b6040518082815260200191505060405180 9
    10390f35b7f4d99b957a2bc29a30ebd96a7be8e68fe50a3c70
    1db28a91436490b7d53870ca43382604051808373fffffffff
    fffffffffffffffffffffffffffffff1673fffffffffffffff
    ffffffffffffffffffffffffff1681526020018281526020019
    2505050604051809103 90a16000806000 3373fffffffffffff
    fffffffffffffffffffffffffff1673fffffffffffffffffff
    fffffffffffffffffffff16815260200190815260200160002
    05414151561021257600080fd5b6000811180156102225750 6
    00381105b151561022d57600080fd5b806000803373fffffff
    fffffffffffffffffffffffffffffffff1673fffffffffffff
    fffffffffffffffffffffffffff16815260200190815260200
    1600020819055505 0565b61027b610333565b600180546001 8
    160011615610100020316600290048060 1f01602080910402 6
    0200160405190810160405280929190818152602001828054 6
    00181600116156101000203166002900480156103115780601
    f106102e6576101008083540402835291602001916103115 65
    b8201919060005260206000209 05b8154815290600101906 02
    0018083116102f457829003601f168201915b505050505090 5
    090565b6000602052806000526040600020600091509050548
    1565b6020604051908101604052806000815250905600a1656
    27a7a72305820ec7d3e1dae8412ec85045a8eafc248e37ae50
    6802cc008ead300df1ac81aab490029'
})
```

```
        .send({
                from: '0xbaf735f889d603f0ec6b1030c91d9033e6
                0525c3',
                gas: 4700000,
                gasPrice: '20000000000000'
            },
            function (error, transactionHash) {
                console.log(error);
                console.log(transactionHash);
            })
        .then(function (contract) {
            console.log(contract);
        });
};
```

After executing this function from our node.js application, we received the following output:

```
Contract {
  currentProvider: [Getter/Setter],
  _requestManager:
   RequestManager {
     provider: null,
     providers:
      { WebsocketProvider: [Function: WebsocketProvider],
        HttpProvider: [Function: HttpProvider],
        IpcProvider: [Function: IpcProvider] },
     subscriptions: {} },
  givenProvider: null,
  providers:
   { WebsocketProvider: [Function: WebsocketProvider],
     HttpProvider: [Function: HttpProvider],
     IpcProvider: [Function: IpcProvider] },
```

```
_provider: null,
setProvider: [Function],
BatchRequest: [Function: bound Batch],
extend:
 { [Function: ex]
   formatters:
    { inputDefaultBlockNumberFormatter: [Function:
    inputDefaultBlockNumberFormatter],
      inputBlockNumberFormatter: [Function:
      inputBlockNumberFormatter],
      inputCallFormatter: [Function: inputCallFormatter],
      inputTransactionFormatter: [Function:
      inputTransactionFormatter],
      inputAddressFormatter: [Function:
      inputAddressFormatter],
      inputPostFormatter: [Function: inputPostFormatter],
      inputLogFormatter: [Function: inputLogFormatter],
      inputSignFormatter: [Function: inputSignFormatter],
      outputBigNumberFormatter: [Function:
      outputBigNumberFormatter],
      outputTransactionFormatter: [Function:
      outputTransactionFormatter],
      outputTransactionReceiptFormatter: [Function:
      outputTransactionReceiptFormatter],
      outputBlockFormatter: [Function: outputBlockFormatter],
      outputLogFormatter: [Function: outputLogFormatter],
      outputPostFormatter: [Function: outputPostFormatter],
      outputSyncingFormatter: [Function:
      outputSyncingFormatter] },
```

```
utils:
 { _fireError: [Function: _fireError],
   _jsonInterfaceMethodToString: [Function:
   _jsonInterfaceMethodToString],
   randomHex: [Function: randomHex],
   _: [Function],
   BN: [Function],
   isBN: [Function: isBN],
   isBigNumber: [Function: isBigNumber],
   isHex: [Function: isHex],
   isHexStrict: [Function: isHexStrict],
   sha3: [Function],
   keccak256: [Function],
   soliditySha3: [Function: soliditySha3],
   isAddress: [Function: isAddress],
   checkAddressChecksum: [Function: checkAddressChecksum],
   toChecksumAddress: [Function: toChecksumAddress],
   toHex: [Function: toHex],
   toBN: [Function: toBN],
   bytesToHex: [Function: bytesToHex],
   hexToBytes: [Function: hexToBytes],
   hexToNumberString: [Function: hexToNumberString],
   hexToNumber: [Function: hexToNumber],
   toDecimal: [Function: hexToNumber],
   numberToHex: [Function: numberToHex],
   fromDecimal: [Function: numberToHex],
   hexToUtf8: [Function: hexToUtf8],
   hexToString: [Function: hexToUtf8],
   toUtf8: [Function: hexToUtf8],
   utf8ToHex: [Function: utf8ToHex],
   stringToHex: [Function: utf8ToHex],
```

```
        fromUtf8: [Function: utf8ToHex],
        hexToAscii: [Function: hexToAscii],
        toAscii: [Function: hexToAscii],
        asciiToHex: [Function: asciiToHex],
        fromAscii: [Function: asciiToHex],
        unitMap: [Object],
        toWei: [Function: toWei],
        fromWei: [Function: fromWei],
        padLeft: [Function: leftPad],
        leftPad: [Function: leftPad],
        padRight: [Function: rightPad],
        rightPad: [Function: rightPad],
        toTwosComplement: [Function: toTwosComplement] },
     Method: [Function: Method] },
  clearSubscriptions: [Function],
  options:
   { address: [Getter/Setter],
     jsonInterface: [Getter/Setter],
     data: undefined,
     from: undefined,
     gasPrice: undefined,
     gas: undefined },
  defaultAccount: [Getter/Setter],
  defaultBlock: [Getter/Setter],
  methods:
   { votes: [Function: bound _createTxObject],
     '0xd8bff5a5': [Function: bound _createTxObject],
     'votes(address)': [Function: bound _createTxObject],
     getPoll: [Function: bound _createTxObject],
     '0x03c32278': [Function: bound _createTxObject],
     'getPoll()': [Function: bound _createTxObject],
```

```
      vote: [Function: bound _createTxObject],
      '0x0121b93f': [Function: bound _createTxObject],
      'vote(uint256)': [Function: bound _createTxObject] },
   events:
    { Voted: [Function: bound ],
      '0x4d99b957a2bc29a30ebd96a7be8e68fe50a3c701db28a91436
      490b7d53870ca4': [Function: bound ],
      'Voted(address,uint256)': [Function: bound ],
      allEvents: [Function: bound ] },
   _address: '0x59E7161646C3436DFdF5eBE617B4A172974B481e',
   _jsonInterface:
    [ { constant: true,
        inputs: [Array],
        name: 'votes',
        outputs: [Array],
        payable: false,
        stateMutability: 'view',
        type: 'function',
        signature: '0xd8bff5a5' },
      { constant: true,
        inputs: [],
        name: 'getPoll',
        outputs: [Array],
        payable: false,
        stateMutability: 'view',
        type: 'function',
        signature: '0x03c32278' },
      { anonymous: false,
        inputs: [Array],
        name: 'Voted',
        type: 'event',
```

```
    signature: '0x4d99b957a2bc29a30ebd96a7be8e68fe50a3c70
    1db28a91436490b7d53870ca4' },
  { constant: false,
    inputs: [Array],
    name: 'vote',
    outputs: [],
    payable: false,
    stateMutability: 'nonpayable',
    type: 'function',
    signature: '0x0121b93f' } ] }
```

The output shows the various properties of the contract we deployed to our private network. The most important one is the contract address at which the contract is deployed, which is **0x59E7161646C3436DFdF5eBE617B4A172974B481e**.

The contract ABI and address can be used to call a function on the contract. In the next section we will build a simple web app that will call the vote function of this contract, showcasing how the polling can be done from the front end.

Client Application

As we did in the last chapter, we can use the web3 library to call a function on a smart contract. But, in the last chapter we did that using a node.js application and not in a browser application. In this section, we will be using web3 in a browser application to call our deployed smart contract's vote function.

The simplest web application we can create for this DApp is a single web page with a few text and button controls. For the web page, we can use the following code inside an html file and then run it from a local server. Note that running from a local server and not directly opening the file from the browser is important to load the scripts properly, without facing any browser security issues.

```
<html>

<c>
    <meta charset="UTF-8">
    <title>Beginning Blockchain - DApp demo</title>
    <script src="<source of web3 library from any CDN or local
    file>"></script>
</head>

<body>
    <div>
        <p>
            <strong>Beginning Blockchain</strong>
        </p>
        <p>Hi, Welcome to the Polling DApp!</p>
        <p> </p>
        <p>Get latest poll: 
            <button onclick="getPoll()">Get Poll</button>
        </p>
        <p>
            <div id="pollSubject"></div>
        </p>
        <p>Vote: Yes:
            <input type="radio" id="yes"> No:
            <input type="radio" id="no">
        </p>
        <p>Submit: 
            <button onclick="submitVote()">Submit Vote</button>
        </p>
        </p>
    </div>
```

```
<script>
    if (typeof web3 !== 'undefined') {
        web3 = new Web3(web3.currentProvider);
    } else {
        web3 = new Web3(new Web3.providers.HttpProvider
        ('http://127.0.0.1:8507'));
    }

    function getPoll() {
        var pollingContract = new web3.eth.Contract([{
                "constant": true,
                "inputs": [{
                    "name": "",
                    "type": "address"
                }],
                "name": "votes",
                "outputs": [{
                    "name": "",
                    "type": "uint256"
                }],
                "payable": false,
                "stateMutability": "view",
                "type": "function"
            },
            {
                "constant": true,
                "inputs": [],
                "name": "getPoll",
                "outputs": [{
                    "name": "",
                    "type": "string"
                }],
```

```
        "payable": false,
        "stateMutability": "view",
        "type": "function"
    },
    {
        "anonymous": false,
        "inputs": [{
                "indexed": false,
                "name": "_voter",
                "type": "address"
            },
            {
                "indexed": false,
                "name": "_value",
                "type": "uint256"
            }
        ],
        "name": "Voted",
        "type": "event"
    },
    {
        "constant": false,
        "inputs": [{
            "name": "selection",
            "type": "uint256"
        }],
        "name": "vote",
        "outputs": [],
        "payable": false,
        "stateMutability": "nonpayable",
        "type": "function"
    }
```

```
        ], '0x59E7161646C3436DFdF5eBE617B4A172974B481e');

        pollingContract.methods.getPoll().call().
        then(function (value) {
            document.getElementById('pollSubject').
            textContent = value;
        });
    };

    function submitVote() {
        var value = 0
        var yes = document.getElementById('yes').checked;
        var no = document.getElementById('no').checked;

        if (yes) {
            value = 1
        } else if (no) {
            value = 2
        } else {
            return;
        }

        var pollingContract = new web3.eth.Contract([{
                "constant": true,
                "inputs": [{
                    "name": "",
                    "type": "address"
                }],
                "name": "votes",
                "outputs": [{
                    "name": "",
                    "type": "uint256"
                }],
```

```
            "payable": false,
            "stateMutability": "view",
            "type": "function"
        },
        {
            "constant": true,
            "inputs": [],
            "name": "getPoll",
            "outputs": [{
                "name": "",
                "type": "string"
            }],
            "payable": false,
            "stateMutability": "view",
            "type": "function"
        },
        {
            "anonymous": false,
            "inputs": [{
                    "indexed": false,
                    "name": "_voter",
                    "type": "address"
                },
                {
                    "indexed": false,
                    "name": "_value",
                    "type": "uint256"
                }
            ],
            "name": "Voted",
            "type": "event"
        },
```

```
                {
                    "constant": false,
                    "inputs": [{
                        "name": "selection",
                        "type": "uint256"
                    }],
                    "name": "vote",
                    "outputs": [],
                    "payable": false,
                    "stateMutability": "nonpayable",
                    "type": "function"
                }
            ], '0x59E7161646C3436DFdF5eBE617B4A172974B481e');

            pollingContract.methods.vote(value).send({
                from: '0xbaf735f889d603f0ec6b1030c91d9033e
                60525c3'
            }).then(function (result) {
                console.log(result);
            });
        };
    </script>
</body>

</html>
```

Let's now analyze each of the sections of this HTML file.

In the head section of the HTML document, we have loaded the web3 script from either a CDN source or a local source. This is just like we refer to any other third-party JavaScript library in our web pages (JQuery, etc.)

Then in the body section of the HTML, we have the controls for showing the poll subject and radio and submit buttons to capture user input. The overall web page looks like this (Figure 6-11).

Beginning Blockchain

Hi, Welcome to the Polling DApp!

Get latest poll: Get Poll

Vote: Yes: No:

Submit: Submit Vote

Figure 6-11. *Polling web application view*

What is important is the script section in the body. That's where we are calling the smart contract interaction code. Let's look at it in detail.

```
<script>
        if (typeof web3 !== 'undefined') {
            web3 = new Web3(web3.currentProvider);
        } else {
            web3 = new Web3(new Web3.providers.HttpProvider
            ('http://127.0.0.1:8507'));
        }

        function getPoll() {
            var pollingContract = new web3.eth.Contract([{
                    "constant": true,
                    "inputs": [{
                        "name": "",
                        "type": "address"
                    }],
```

```
        "name": "votes",
        "outputs": [{
            "name": "",
            "type": "uint256"
        }],
        "payable": false,
        "stateMutability": "view",
        "type": "function"
    },
    {
        "constant": true,
        "inputs": [],
        "name": "getPoll",
        "outputs": [{
            "name": "",
            "type": "string"
        }],
        "payable": false,
        "stateMutability": "view",
        "type": "function"
    },
    {
        "anonymous": false,
        "inputs": [{
                "indexed": false,
                "name": "_voter",
                "type": "address"
            },
```

```
                {
                    "indexed": false,
                    "name": "_value",
                    "type": "uint256"
                }
            ],
            "name": "Voted",
            "type": "event"
        },
        {
            "constant": false,
            "inputs": [{
                "name": "selection",
                "type": "uint256"
            }],
            "name": "vote",
            "outputs": [],
            "payable": false,
            "stateMutability": "nonpayable",
            "type": "function"
        }
    ], '0x59E7161646C3436DFdF5eBE617B4A172974B481e');

    pollingContract.methods.getPoll().call().
    then(function (value) {
        document.getElementById('pollSubject').
        textContent = value;
    });
};
```

```
function submitVote() {
    var value = 0
    var yes = document.getElementById('yes').checked;
    var no = document.getElementById('no').checked;

    if (yes) {
        value = 1
    } else if (no) {
        value = 2
    } else {
        return;
    }

    var pollingContract = new web3.eth.Contract([{
            "constant": true,
            "inputs": [{
                "name": "",
                "type": "address"
            }],
            "name": "votes",
            "outputs": [{
                "name": "",
                "type": "uint256"
            }],
            "payable": false,
            "stateMutability": "view",
            "type": "function"
        },
        {
            "constant": true,
            "inputs": [],
            "name": "getPoll",
```

```
            "outputs": [{
                "name": "",
                "type": "string"
            }],
            "payable": false,
            "stateMutability": "view",
            "type": "function"
        },
        {
            "anonymous": false,
            "inputs": [{
                    "indexed": false,
                    "name": "_voter",
                    "type": "address"
                },
                {
                    "indexed": false,
                    "name": "_value",
                    "type": "uint256"
                }
            ],
            "name": "Voted",
            "type": "event"
        },
        {
            "constant": false,
            "inputs": [{
                "name": "selection",
                "type": "uint256"
            }],
            "name": "vote",
            "outputs": [],
```

```
                "payable": false,
                "stateMutability": "nonpayable",
                "type": "function"
            }
        ], '0x59E7161646C3436DFdF5eBE617B4A172974B481e');

        pollingContract.methods.vote(value).send({
            from: '0xbaf735f889d603f0ec6b1030c91d9033
            e60525c3'
        }).then(function (result) {
            console.log(result);
        });
    };
</script>
```

In the previous script section, first we are initializing the **web3** object with the HTTP provider of the local Ethereum node (if it is not already initialized).

Then, we have two JavaScript functions. One for getting the value of the **pollSubject** string from the smart contract and another for calling the **vote** function of the contract.

The calling of smart contract functions is exactly how we did it in the previous chapter using the **web3.eth.Contract** submodule of the **web3** library.

Note that in the first function **getPoll** we are calling the **call** function on the smart contract instance, while in the second function **submitVote** we are calling **send** on the smart contract instance. That's primarily the difference in the two function calls.

Using the **call** on the **getPoll** function of the smart contract, we are getting the return value of the getPoll function without sending any transaction to the network. We are then showing this value on the UI by assigning it as the text of a UI element.

Next, using **send** on the **vote** function, we are sending a transaction to execute this function on the network and so we have to also define an account that will be used to execute the smart contract function. Following is the output obtained from the submitVote function shown previously, which is basically a transaction receipt.

```
{
        blockHash: '0x04a02dd56c037569eb6abe25e003a65d3366407
        134c90a056f64b62c2d23eb84',
        blockNumber: 4257,
        contractAddress: null,
        cumulativeGasUsed: 43463,
        from: '0xbaf735f889d603f0ec6b1030c91d9033e60525c3',
        gasUsed: 43463,
        logsBloom: '0x000000000000000000000000000000800000000
        00000400000000000000000000200000000000000000000000000
        00000000000000000000000000000000000000000000000000000
        00000000000200000000002000000000000000000000000000000
        00000000000000000000000000000000000000000000000000000
        00000000000000000000000000000000000000000000000000000
        00000000000000000000000000000000000002000000000000000
        00000000000000000000000000000000000000000000000000000
        00000000000000000000000000000000000000000000000000000
        0000000000000000000000000000000000000000000000000',
        root: '0x58bc4ee0a3025ca3f303df9bb243d052a123026519637
        30c52c88aafe92ebeee',
        to: '0x59e7161646c3436dfdf5ebe617b4a172974b481e',
        transactionHash: '0x434aa9c0037af3367a0d3d92985781c50
        774241ace1d382a8723985efcea73b3',
        transactionIndex: 0,
```

```
        events: {
            Voted: {
                address: '0x59E7161646C3436DFdF5eBE617B4A17
                2974B481e',
                blockNumber: 4257,
                transactionHash: '0x434aa9c0037af3367a0d3d929
                85781c50774241ace1d382a8723985efcea73b3',
                transactionIndex: 0,
                blockHash: '0x04a02dd56c037569eb6abe25e003a65
                d3366407134c90a056f64b62c2d23eb84',
                logIndex: 0,
                removed: false,
                id: 'log_980a1744',
                returnValues: [Result],
                event: 'Voted',
                signature: '0x4d99b957a2bc29a30ebd96a7be8e68f
                e50a3c701db28a91436490b7d53870ca4',
                raw: [Object]
            }
        }
}
```

If we look closely at this output, we see that this also has an events section and it shows the triggering of the **Voted** event that we created in our smart contract.

```
events: {
            Voted: {
                address: '0x59E7161646C3436DFdF5eBE617B4A172
                974B481e',
                blockNumber: 4257,
                transactionHash: '0x434aa9c0037af3367a0d3d929
                85781c50774241ace1d382a8723985efcea73b3',
```

```
                    transactionIndex: 0,
                    blockHash: '0x04a02dd56c037569eb6abe25e003a6
                    5d3366407134c90a056f64b62c2d23eb84',
                    logIndex: 0,
                    removed: false,
                    id: 'log_980a1744',
                    returnValues: [Result],
                    event: 'Voted',
                    signature: '0x4d99b957a2bc29a30ebd96a7be8e68fe
                    50a3c701db28a91436490b7d53870ca4',
                    raw: [Object]
        }
    }
```

In the preceding code snippet, we've extracted out the events section from the transaction receipt we received in the response of the send transaction to the vote function of our smart contract. As we can see, the events section also shows the return values and the raw values from the function call.

We have now come to the end of our DApp programming exercise. In the previous sections of this chapter, we developed an end-to-end decentralized application on the Ethereum blockchain and we also deployed a private blockchain for our DApp.

The DApp can be used with the public Ethereum network too–a voter has to host a node and they can vote using their existing Ethereum accounts on the public (main) network.

There can be several ways in which the business logic in the smart contract can be enhanced by using different checks and rules.

This programming exercise gives us a basic idea about how to approach development of decentralized applications and the components that come into the picture during the process. This exercise can be treated as a starting point for Ethereum application development, and the reader is encouraged to explore best practices and more complex scenarios on the subject.

Summary

In this chapter we compiled a programming exercise of developing a decentralized application based on the Ethereum blockchain. We also learned how to set up a private Ethereum network and how to interact with it using the DApp.

References

web3.js Documentation

`http://web3js.readthedocs.io/en/1.0/index.html`.

Solidity Documentation

`https://solidity.readthedocs.org/`.

Ethereum Private Networking Tutorial

`https://github.com/ethereumproject/go-ethereum/wiki/Private-Networking-Tutorial`.

Index

A

Abstraction layers, 17
admin.addPeer() command, 331
Advanced Encryption Standard (AES)
  AddRoundKey, 52
  AES-128, 48
  8-bit byte, 49
  block size, 48
  encryption and decryption process, 49–50
  key expansion, 53–54
  MixColumns, 51
  NIST, 54
  processing steps, 48
  round function, 50
  ShiftRows, 51
  state array, 49
  state words, 49
  SubBytes, 50
  substitution-permutation network, 48–49
Amazon, 14, 23, 184
Application Binary Interface (ABI), 262–263, 297, 302, 308, 310, 336, 337
Application layer, 19, 25
Asymmetric key cryptography
  App stores, 80
  authentication, 79
  code examples, 95–97
  confidentiality, 79
  digital signatures, 78
  DSA, 86–88
  ECC (*see* Elliptic curve cryptography (ECC))
  ECDSA, 93–95
  private key, 80
  public key, 79–81
  RSA algorithm
    encryption/decryption, 84–86
    generation of key pairs, 82–84
    modular arithmetic, 82
  vs. symmetric key cryptography, 102–104
  text message, Alice to Bob, 78
Autonomous Decentralized Peer-To-Peer Telemetry (ADEPT), 27

B

Banking era, 152
Bitcoin, 3, 149
  bitcoinjs, 215, 272
  block explorer API, 272

B. Singhal et al., *Beginning Blockchain*, https://doi.org/10.1007/978-1-4842-3444-0

Bitcoin (*cont.*)
- block structure
  - difficulty target, 165–168
  - header components, 162–163
  - field and size, 161
  - Merkle trees, 163–165
- data structure, 159
- dawn, 153–154
- defined, 154–157
- Ethereum, 220–221
- full nodes, 209–210
- genesis block
  - chainparams.cpp, 169
  - hash information, 171
  - transaction information, 170
- mining, 22
- orphan blocks, 160
- PoW, 22
- smart contracts, 20
- SPVs, 210, 212
- transaction, Bitcoin test network
  - addOutput method, 279
  - broadcast transaction, 281–282
  - transaction.addInput method, 279
  - get test net Bitcoins, 275
  - hex string, 280
  - keypairs creation, 274
  - sender's unspent outputs, 276–278
  - setup and initialization, bitcoinjs library, 273–274
  - sign transaction inputs, 280
- wallets, 212–215
- working with, 157–158

Bitcoin network
- block propagation, 193–194
- consensus and block mining (*see* Block mining)
- discovery, new node, 174–178
- full/lightweight nodes, 173
- on Internet, 172
- SPV, 173–174
- transactions, 179–184

Bitcoin scripts
- CheckSig, 207
- defined, 204
- formation of combined validation, 205
- granular components, 200
- input and output code, 203–204
- practical example, 202
- ScriptPubKey, 201, 203, 207
- ScriptSig, 201
- stack-based implementation, 206–207
- transaction fields, 199
- transactions revisited, 196–198

BitcoinJ, 215

Blockchain
- advantages, 1
- applications
  - actors, handle requests, 137
  - backend database, 135
  - Bitcoin node, 135–136
  - centralized web server, 135

cloud-empowered blockchain system, 136–137
cloud services, 136, 138
consensus algorithms, 137
DApps, Ethereum network, 138
development, 269–270
hybrid, 138
interaction, 271
public blockchain, 136
banking system, 2–3
Bitcoin (*see* Bitcoin)
business problems and situations, 34
Byzantine Generals' Problem, 33
centralized system, 33
components, 32
computer science engineering (*see* Computer science engineering)
core, 32
cryptocurrency implementations, 32
cryptography (*see* Cryptography)
data structure, 9, 123
decentralized and peer-to-peer solution, 6–7
description, 31
distributed consensus mechanisms, 130–131
fundamentals, 122
game theory (*see* Game theory)
handcrafting transactions, 312–313
intermediary *vs.* peer-to-peer transaction, 4–5
offerings, 24
PBFT, 134–135
PoS, 133–134
PoW, 131–133
properties
auditability, 127
consistent state of ledger, 127
democratic, 125
double-spend resistant, 126
forgery resistant, 125
immutability, 125
resilient, 127
real-world business problems, 31
scalability
Bitcoin adoption, 139
centralized system, 139
consensus protocols, 139
database sharding, 143–145
disruptive technologies, 139
off-chain computation, 140–143
public and private Blockchains, 140
transactions, 139
scenarios, 123
transactions, 127–129, 312
use cases, 26–27

Block ciphers, 40–41
Block mining
ballpark values, 190
block header, 189
block reward, 187
coin creation, 187
cryptographic security, 185
defined, 184
halving process, 187
hash and target value, 190
incentivization mechanism, 192
miners, 188
nodes, 188
orphaned blocks, 193
PoW, 185, 191
transaction fees, 186
valid block, 190
Bureaucratic system, 3
Business transaction, 4
Byzantine Generals' Problem, 110–112, 114

C

Centralized systems
advantages, 14
vs. decentralized systems, 11–14
limitations, 14, 23–24
Coinbase transaction, 179
Computer science engineering
blockchain
block-1234, 116
block structure, 117
data structure, 114
genesis block, 115
hash pointer, 114–115
parent block, 115
SHA-256, 116
Merkle trees, 117–122
Consensus layer, 22
Contract.deploy method, 303
Cryptography
advanced mathematical techniques, 34
asymmetric key (*see* Asymmetric key cryptography)
authentication, 35
ciphertext, 35–36
confidentiality, 35
data integrity, 35
Diffie-Hellman key exchange, 98–101
encryption techniques, 35
hash functions (*see* Hash functions)
non-repudiation, 35
plaintext, 35
steps, 36
symmetric key (*see* Symmetric key cryptography)
transactions, 33

D

Database sharding, 143–145
Data Encryption Standard (DES)
64-bit block size, 43

cryptography, 44
Feistel cipher, 43, 45–47
key generator, 44
limitations, 48
Moore's law, 43
round function, 47
Decentralized applications (DApps)
architecture
public nodes *vs.* self-hosted nodes, 315–316
servers, 316
blockchain-based, 268
client application, web3
getPoll function, 371
html file and scripts, 360
JavaScript functions, 371
polling web application view, 366
send on vote function, 372
smart contract interaction code, 366
transaction, smart contract function, 372–373
voted event, 373–374
web3.eth.Contract submodule, 371
private Ethereum network (*see* Private Ethereum network)
smart contract (*see* Smart contract, DApp)
voting system, 269
Decentralized applications (DApps), 138
Decentralized systems
advantages, 15, 24
vs. centralized systems, 11–14
limitations, 15
peer-to-peer system, 16
Diffie-Hellman key exchange, 98–101
Digital signature algorithm (DSA), 62, 86–88

E

Elliptic curve cryptography (ECC)
160-bit ECC key, 88
characteristics, 89–92
discrete logarithm problem, 88
domain parameters, 92
mathematical equation, 88
shapes, 88
Elliptic Curve Diffie-Hellman (ECDH), 93
Elliptic Curve Digital Signature Algorithm (ECDSA)
key generation, 93
vs. RSA, 93
sender and receiver, 93
signature generation, 94
signature verification, 94–95
Ethereum blockchain, 219
accounts
advantages, 232–233
Contract Accounts, 228
EOAs (*see* Externally Owned Accounts (EOAs))

Ethereum blockchain (*cont.*)
state, 233–235
UTXOs advantages, 231–232
Bitcoin to, 220–221
block metadata, 226
consensus-subsystem information, 227
data references, 226
data structure, 225
decentralized applications, 221, 222
design philosophy, 223–224
ecosystem
DApp, 264
development components, 265
limitations, 263
Swarm, 264
Whisper, 264
EVM, 222, 257–262
gas and transaction cost, 248–253
Infura API service, 284
Merkle Patricia tree, 237–239
mining, 22
PoW, 22
RLP encoding, 239
Ropsten test network
library and connection, 284–285
preparation, 287–288
send transaction, 290–292
set up Ethereum accounts, 285–286
sign transaction, 288–289
testnet faucets, sender's account, 286–287
smart contracts, 20, 253–254
application, 256
blocks, 255
compilation, 297
contract creation, 256–257
deploy, 302
executing, 309–311
Remix IDE, 294–295
solidity programming language, 293
transaction, 295, 297
voting application, 255
software development and deployment, 223
state transaction function, 245–247
transaction and message structure, 240–244
transaction execution information, 227
trie usage, 236
Ethereum virtual machine (EVM)
ABI, 262–263
absolute determinism, 258
easy security, 258
JVM, 222
memory, 261–262
native operations, 258

P2P network, 259
simplicity, 257–258
smart contract deployment and usage, 259–260
space optimization, 258
stack, 262
storage, 260–261
Ethminer, 253
Execution layer, 20
Externally Owned Accounts (EOAs), 228
to Contract Account transaction, 230–231
to EOA transaction, 229

F

Feistel cipher, 43, 45–47
Fiat currency, 152
Financial services market, 25

G

Game theory
Bitcoins, 104
blockchain job, 104
Byzantine Generals' Problem, 110–112, 114
cricket tournament, 104
Nash Equilibrium, 107–108
prisoner's dilemma, 108–110, 113
real-life situations, 104, 106, 113
sport event, 106
strategies, 105
vegetables, 105
zero-sum games, 112–113
Government sectors, 28

H

Handcrafting transactions, 312–313
Hash functions
applications, 73
basic form, 56
Bitcoin, 60
code examples, 74–75
core properties, 56
hash value, 56
information security applications, 55
message digest (MD) family, 62
puzzle friendliness, 60
RIPEMD, 67
search puzzle, 61
security properties
collision resistance, 57–58
pre-image resistance, 58–60
SHA (*see* Secure Hash Algorithm (SHA))
Hyperledger, 20, 24, 117, 134, 139

I, J

Initial Coin Offering (ICO), 26
Internet Engineering Task Force (IETF), 74

K

Keypairs, 274

L

Layers
abstraction, 17
application, 19
consensus, 22
execution, 20
propagation, 21–22
semantic, 20–21

M

Merkle trees, 21, 117–122
Message authentication code (MAC), 55, 76–77
Mining, 156
Mist wallet, 265
Monetary transactions, 9–11
Money
banking era, 151–152
fiat currency, 152, 153
gold and silver metals, 151
Internet, 153
pimitive barter system, 150

N, O

Nash Equilibrium, 107–108
National Institute of Standards and Technology (NIST), 54
National Security Agency (NSA), 62

P, Q

PBFT, *see* Practical Byzantine Fault Tollerance (PBFT)
Pimitive barter system, 150
PoS, *see* Proof of Stake (PoS)
PoW, *see* Proof of Work (PoW)
Practical Byzantine Fault Tollerance (PBFT), 134–135
Prisoner's dilemma, 108–110, 113
Private Ethereum network
account creation, 323
first node
configuration, 327
custom genesis configuration, 326
geth command, 326–328
genesis.json configuration file, 324–325
geth data directory, 322
install geth, 321
second node
command, 330
genesis.json configuration, 329
geth console, peers, 331–332
geth initialize configuration, 329
geth logs, 333
Proof of Stake (PoS), 133–134
Proof of Work (PoW), 22, 131–133
Propagation layer, 21–22
Pseudorandom number generator (PRNG), 40, 43, 73

Public key infrastructure (PKI), 80–81
Public *vs.* private blockchains, 313–314

R

RACE Integrity Primitives Evaluation Message Digest (RIPEMD), 67
Regular transactions, 179

S

ScriptPubKey, 198
Secure Hash Algorithm (SHA)
- DSA, 62
- NSA, 62
- SHA-1, 62, 64
- SHA-2, 63–64
- SHA-3
  - cryptographic hash functions, 68
  - Merkle-Damgård construction, 68
  - NIST, 67
  - sponge construction, 68–70, 72
  - state array representationin, 71
  - variants, padding, 69
- SHA-256 and SHA-512, 65–66
- versions, 62

Semantic layer, 20–21
sendSignedTransaction function, 290
Simplified Payment Verification (SPV), 173
signTransaction function, 289–290
Smart contract, Ethereum DApp
- client applications
  - html code, 360
  - interaction code, 366
  - polling web application view, 366
  - transaction, 372
- creation
  - ABI, 337, 339
  - byte code, 339
  - polling functionality, 334
  - preceding code snippet, 335
  - Remix online Solidity editor, 337
  - solidity code snippet, 334
  - voting functionality, 336
- deploying
  - private network, 345
  - web3 library and connection, 345

Smart software engineering, 34
Solidity programming language, 293
Stock transaction, 5–6
Stream ciphers, 39–40
Supply chains, 28
Symmetric key cryptography
- AES (*see* Advanced Encryption Standard (AES))

Symmetric key cryptography (*cont.*)
vs. asymmetric key cryptography, 102–104
block ciphers, 40–41
ciphertext, 37
DES (*see* Data Encryption Standard (DES))
file transfer protocols, 38
Kerckhoff's principle and XOR function, 38–39
limitations, 55
MAC and HMAC, 76–77
one-time pad, 42–43
sender and receiver, 37
"shared secret", 37
stream ciphers, 39–40

T, U, V

transaction.sign function, 280
Transmission Control Protocol/Internet Protocol (TCP/IP), 2, 17
Truffle, 265

W, X, Y

Web3.js, 265
The Wisdom of Crowds, 27
World Wide Web (WWW), 2

Z

Zero-sum games, 112–113